BE YOUR OWN WEATHER EXPERT

JANET KELLY

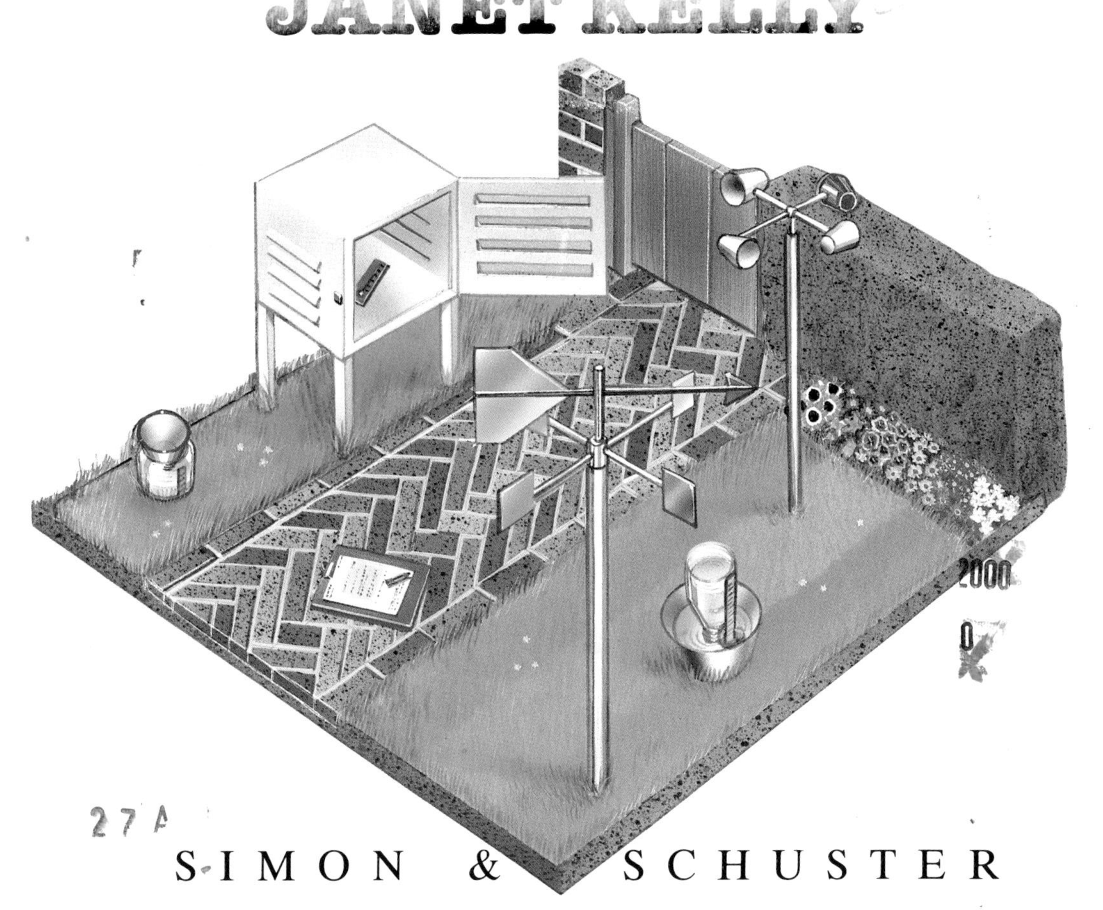

SIMON & SCHUSTER
LONDON • SYDNEY • NEW YORK • TOKYO • SINGAPORE • TORONTO

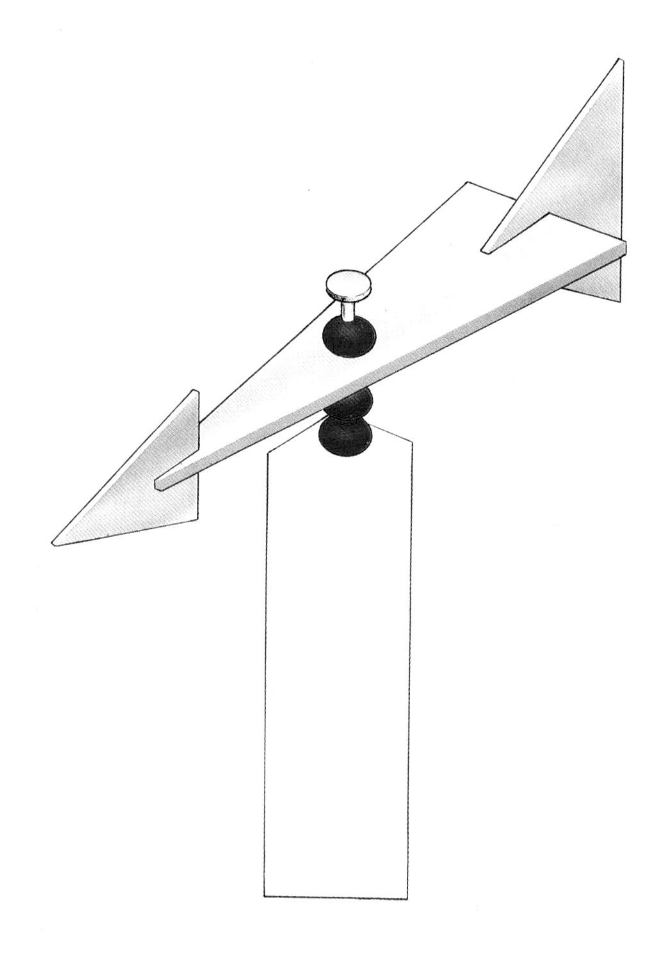

For my father A.G.K.

First published in 1991
by Simon & Schuster Young Books

Simon & Schuster Young Books
Simon & Schuster Ltd
Wolsey House, Wolsey Road,
Hemel Hempstead, Herts HP2 4SS.

Design: David West
Children's Book Design
Editor: Diana Russell
Illustrator: Peter Bull

Printed and bound in Belgium
by Proost International Book Production

A CIP catalogue record for this book is available
from the British Library.

ISBN 0 7500 0847 4

CONTENTS

WHAT IS WEATHER?

Weather is the wind in your face, sunny days, rain, and the cold that numbs your fingers. To scientists, it is "the state of the atmosphere at a particular time and place". Three things are needed to produce weather: the Sun, the air or atmosphere, and water. The usual pattern of weather in a particular place is called the climate.

WHO NEEDS TO KNOW?

Sailors, fishermen, pilots and farmers need to know exactly what weather to expect. If you are planning an outdoor activity, like a picnic or a sport, you will want to know the weather forecast too.

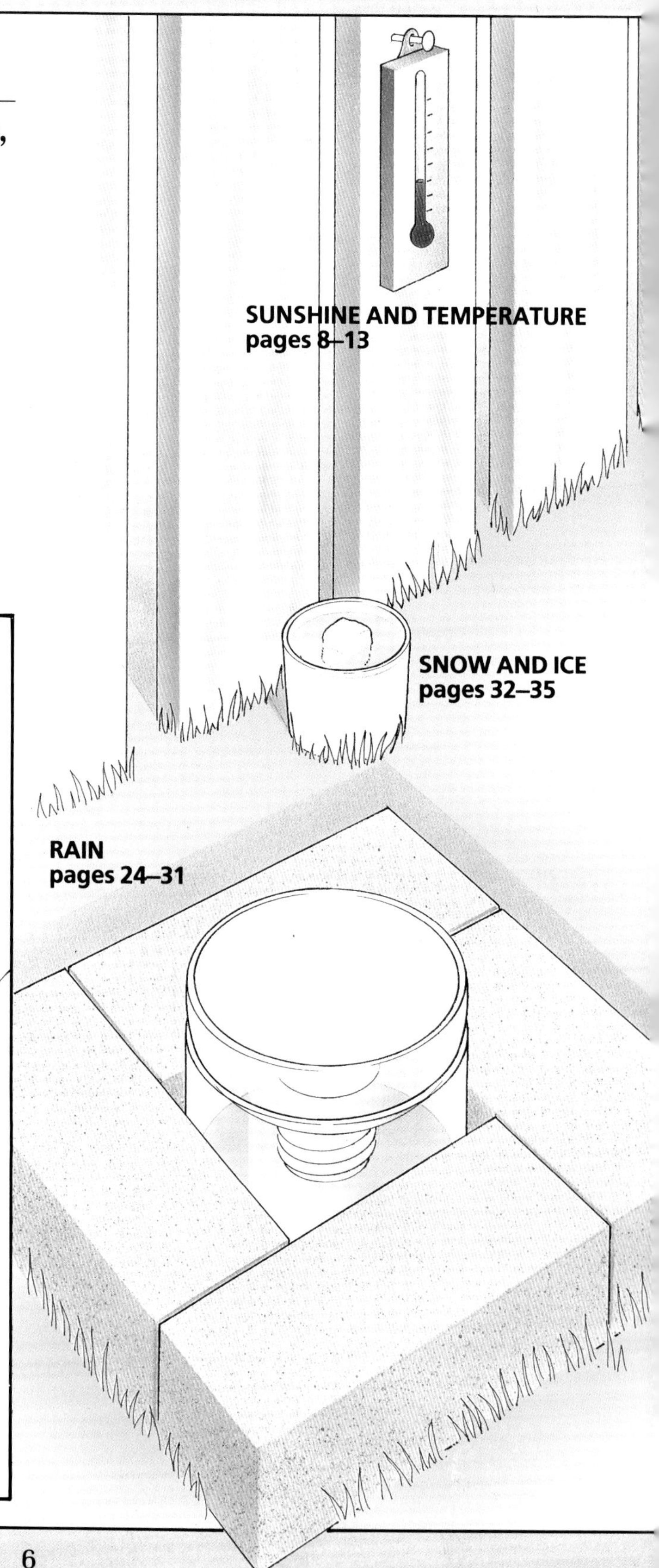

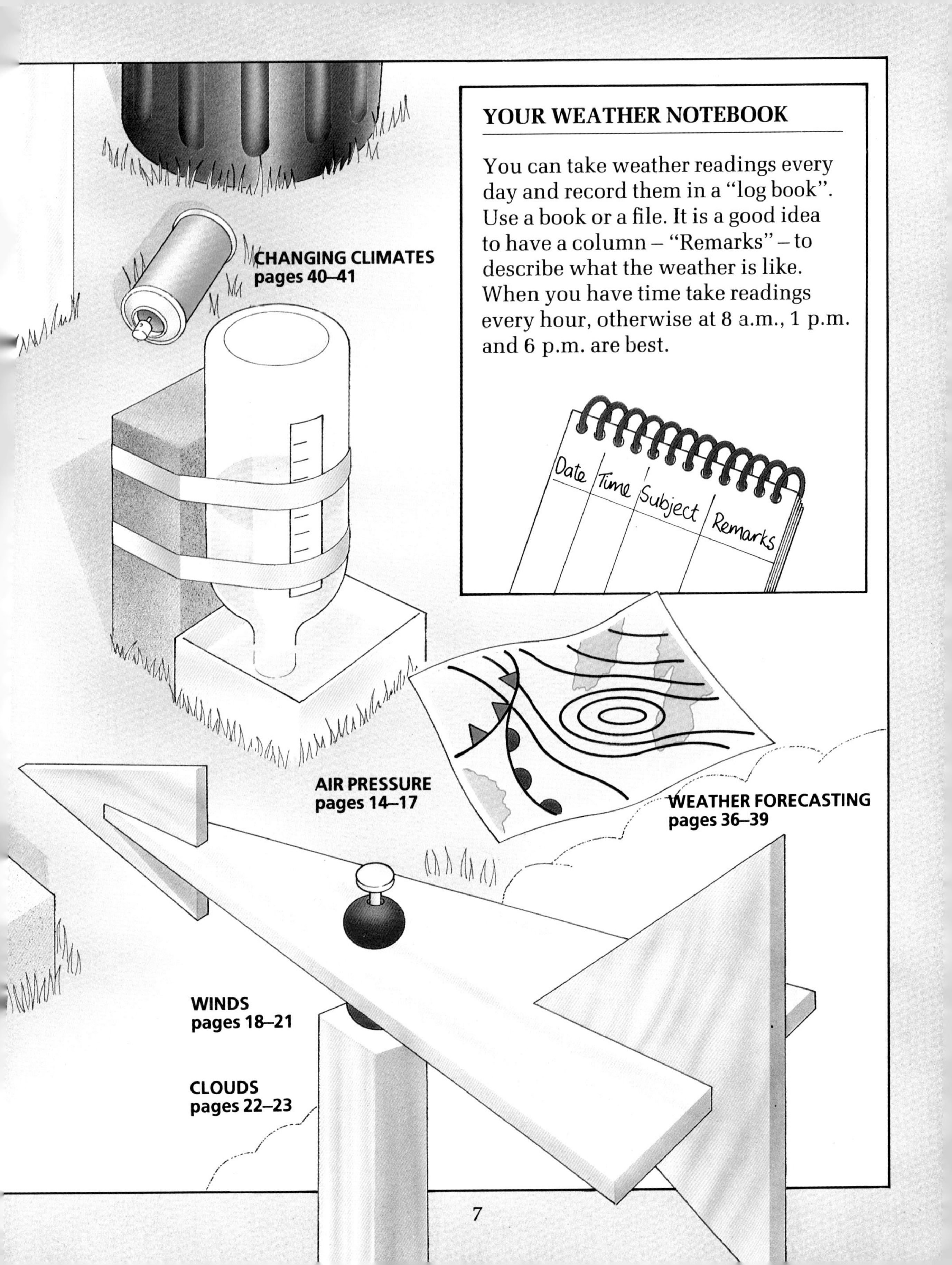

YOUR WEATHER NOTEBOOK

You can take weather readings every day and record them in a "log book". Use a book or a file. It is a good idea to have a column – "Remarks" – to describe what the weather is like. When you have time take readings every hour, otherwise at 8 a.m., 1 p.m. and 6 p.m. are best.

SUNSHINE AND TEMPERATURE

The story of Earth's weather starts 148 million kilometres away with the Sun. The Sun is a huge, hot churning ball of exploding gases, which sends out enormous amounts of energy in the form of heat and light. The Sun's rays pass through the atmosphere and warm the Earth's surface, which then heats the air above it. How hot or cold this air is, is called its temperature, which is measured with thermometers.

DAY AND NIGHT

Day and night occur because Earth spins on its axis like a top, and each complete spin takes 24 hours. When one side of Earth faces the Sun, it is in daylight and the Sun warms the land. When this side spins away from the Sun it goes into darkness and night-time, and the land gradually loses heat.

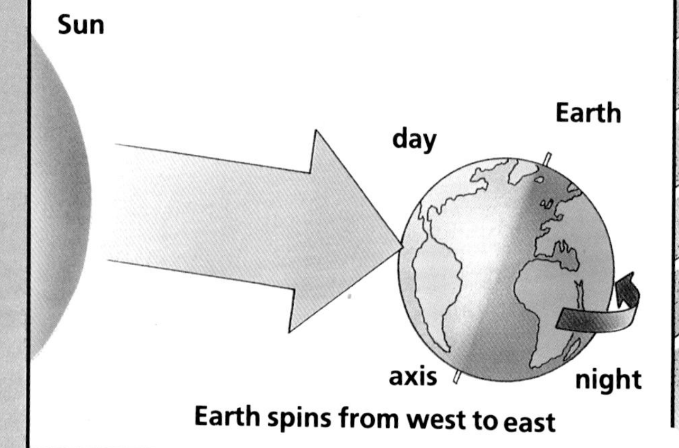

Earth spins from west to east

Fix the thermometer to a north-facing wall in the shade. Make sure you can read it – put it at eye level. Take the temperature every day and record in your notebook.

Date	Time	Temp	Remarks
25 July	9am 12am 6pm	15°C 24°C 18°C	morning mist sunny day warm evening
26 July	8:30am 12am	13°C 15°C	Cloudy starting to rain

MAKING A STEVENSON SCREEN

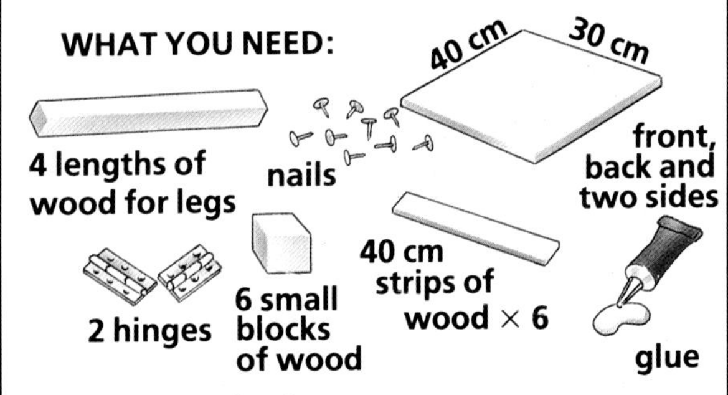

How to make it

Ask an adult to help.

1 Cut the plywood to size, saw off small sections from the sides. Drill holes in the front, side and back pieces of wood.

2 Glue or nail the top, base and sides together, and the legs to the base.

3 Fix hinges.

4 Cut the blocks of wood as shown, glue to the door, and attach the thin strips of wood.

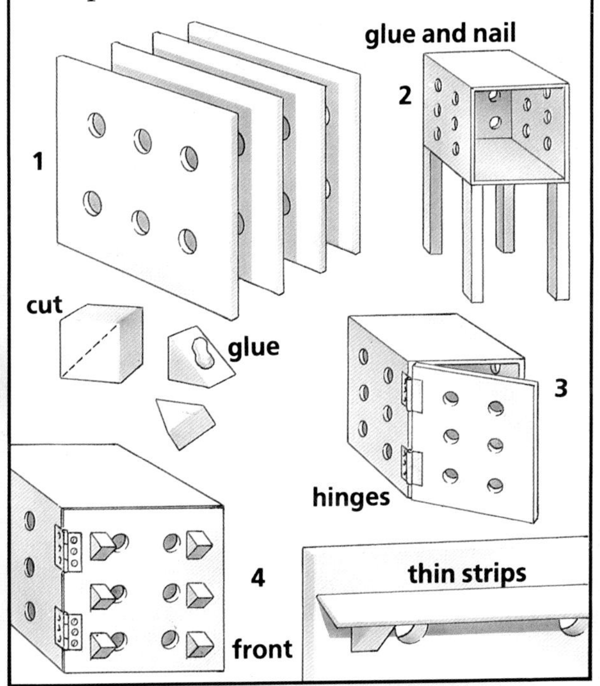

Thermometers are usually placed in a box called a Stevenson screen, where they are protected from sunlight and wind, but where air is allowed to flow freely around them. The screen should be at least 1 metre off the ground and placed on open ground, like a lawn.

THERMOMETERS

Alcohol and mercury thermometers measure temperature. Maximum and minimum thermometers contain alcohol and mercury and record highest and lowest temperatures during a day. The mercury and alcohol move as the temperature changes and they push a metal pin along the tubes. The furthest points reached show the highest and lowest temperatures.

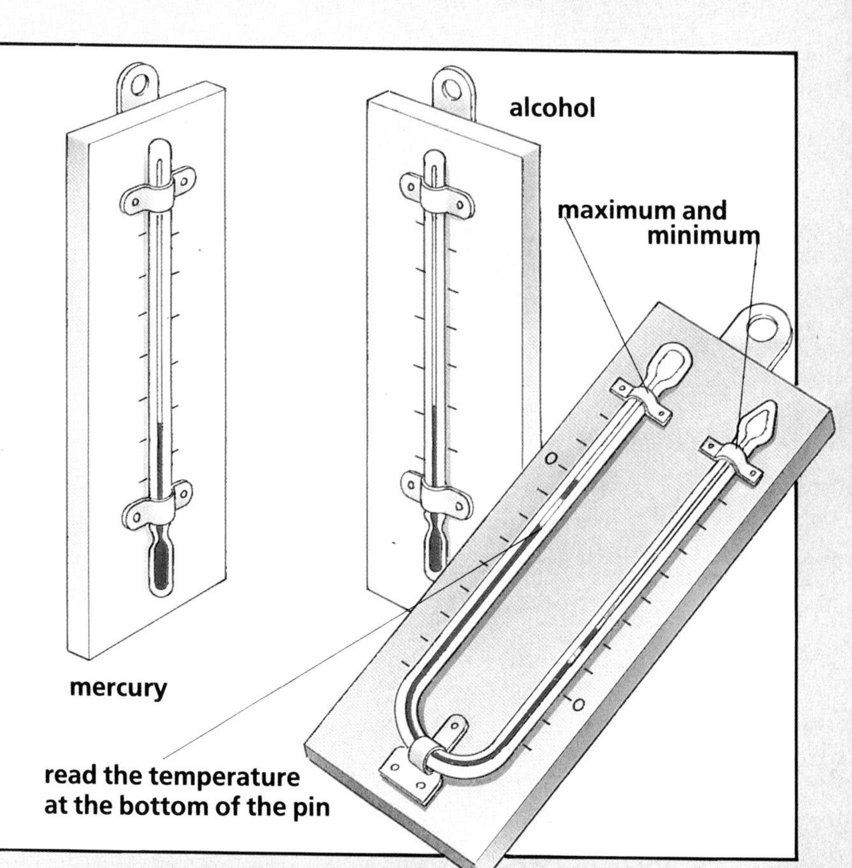

PHOTOSYNTHESIS

Green plants use light energy from the Sun to make carbohydrates (sugars and starches) from carbon dioxide in the air and water from the soil. This is photosynthesis.

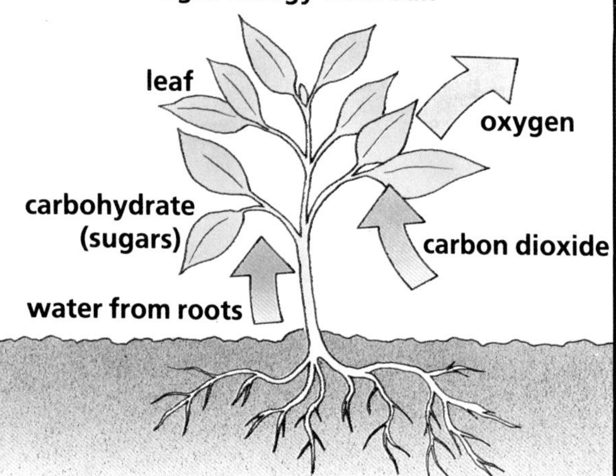

COOL COLOURS

Find two thermometers. Put one under a black, the other under a white sheet of paper. Leave in the shade, read the temperatures later. The one under black paper should be higher. Dark colours absorb heat, light ones reflect it.

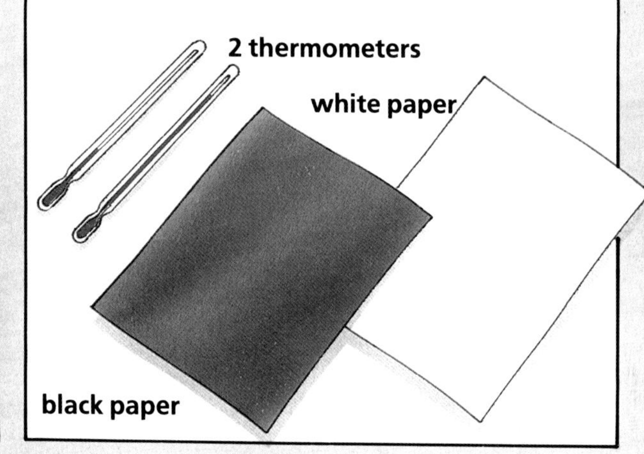

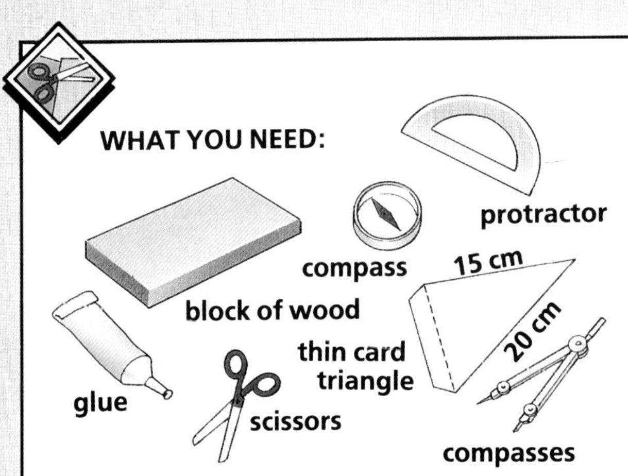

MAKING A SUNDIAL

1 Use a piece of wood or a book for the base. Cover this with white paper and use a pencil and compasses to draw a semi-circle on it.

2 Use a protractor to draw a right-angled triangle and flap on a piece of card. Carefully cut it out, fold it and glue the flap to the base.

3 Place the sundial on flat ground in the sun and use a compass to line up the sundial on a north–south line. The card will cast a shadow on the base.

Making it work

Using a watch, mark and label the shadow position every hour on the semi-circle. Wait for a sunny day and use your sundial to tell the time.

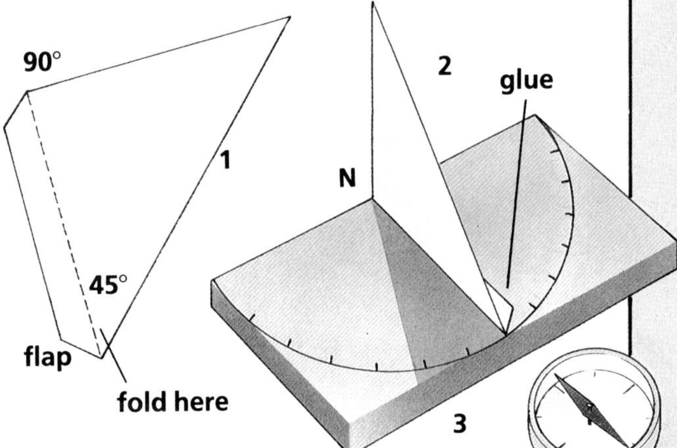

WHAT IS AN ECLIPSE?

Earth and Moon cast their own shadows. An eclipse of the Sun happens when the Moon passes between the Sun and Earth. The Moon's shadow falls on part of Earth, which becomes dark for a time. This is called an eclipse of the Sun.

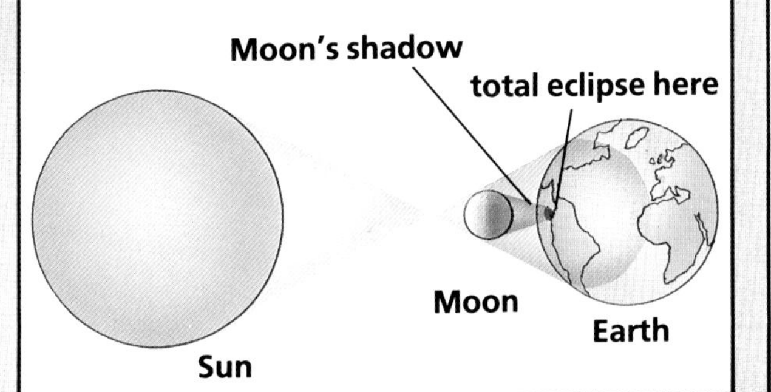

DID YOU KNOW?

The hottest place in the world is Dallol in Ethiopia, where the average shade temperature over a year is 34.4°C.

The coldest place in the world is Vostok in Antarctica, where the average temperature over the year is −57.8°C, and −88°C has been recorded.

To measure sunshine weather forecasters use a Campbell-Stokes recorder. It consists of a glass ball that focuses sunlight onto a piece of card. The sun burns a mark on the card and the length of the mark gives the hours of sunshine.

EARTH'S CLIMATE

Main climate zones are: Polar – at the Arctic and Antarctic there are long dark winters and very short cool summers. Tropical – at the Equator, it is always very hot. Temperate – between the tropics and the poles, there are four seasons: spring, summer (warm and dry), autumn and winter (mild).

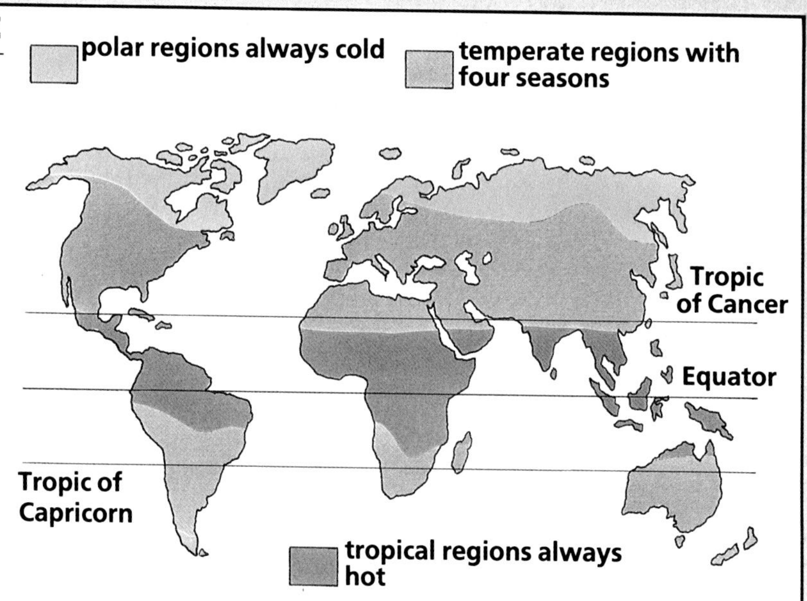

WHY WE HAVE SEASONS

Earth spins on its axis which tilts. When one hemisphere leans towards the sun it gets more heat and light (summer), and when it leans away gets less heat and light (winter).

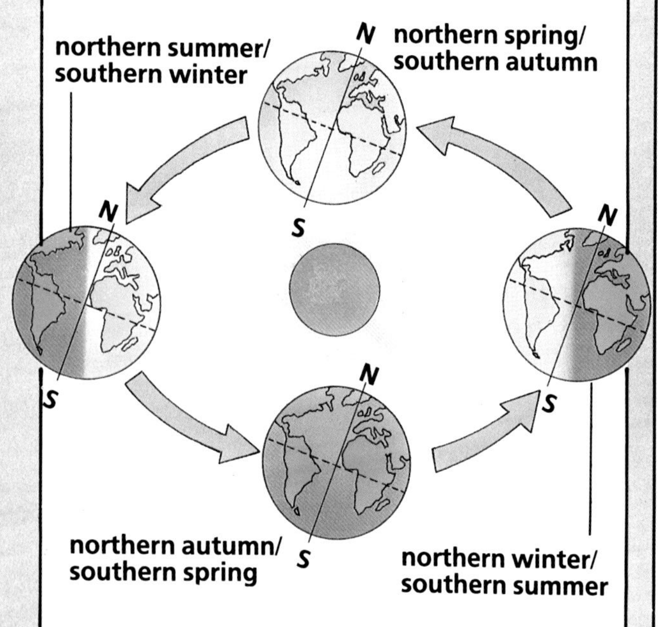

WHY IS IT COLDER AT THE POLES?

The Equator is the hottest part of Earth because the Sun is almost overhead and shines straight on it. The poles are the coldest places because the Earth's surface is curved and so the Sun's rays strike the poles at a slant. The rays spread out and heat a wider area, so they have less heating power.

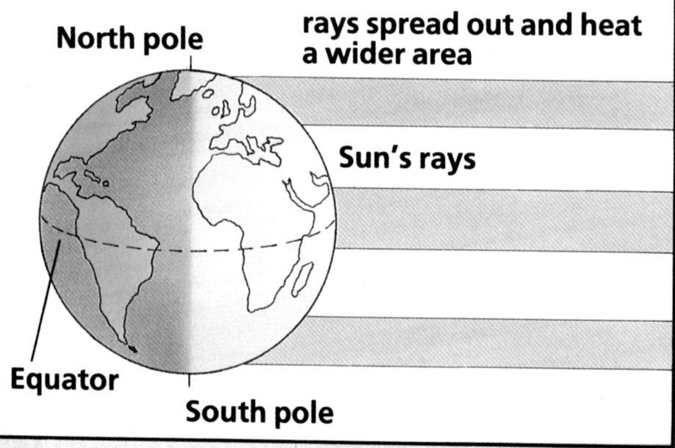

ANIMALS AND THE SEASONS

Animals cope with changing temperatures in many ways. Some hibernate and pass the winter in a deep sleep, while others migrate to warmer countries. Many birds and mammals are active throughout the year and their body temperature remains more or less constant.

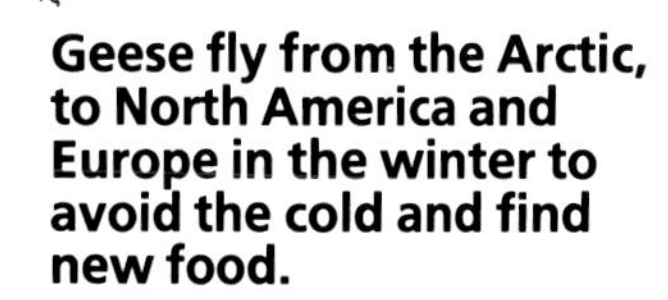

Geese fly from the Arctic, to North America and Europe in the winter to avoid the cold and find new food.

A bird fluffs out its feathers when it is cold, which traps air between the feathers, keeping it warm.

Dogs pant to lose heat. Body heat makes the moisture from their tongues evaporate, which cools them down.

Dormice and hedgehogs hibernate. Their temperature drops and their heartbeat slows right down to save energy.

ARCTIC AND DESERT ANIMALS

Jerboas spend the hot days in burrows, coming out at night to feed. Fennec foxes lose heat through their large ears. Camels store fat in their humps.

The thick coats and small furry ears of polar bears and Arctic foxes, and the fatty blubber under the skins of seals, helps them lose less heat.

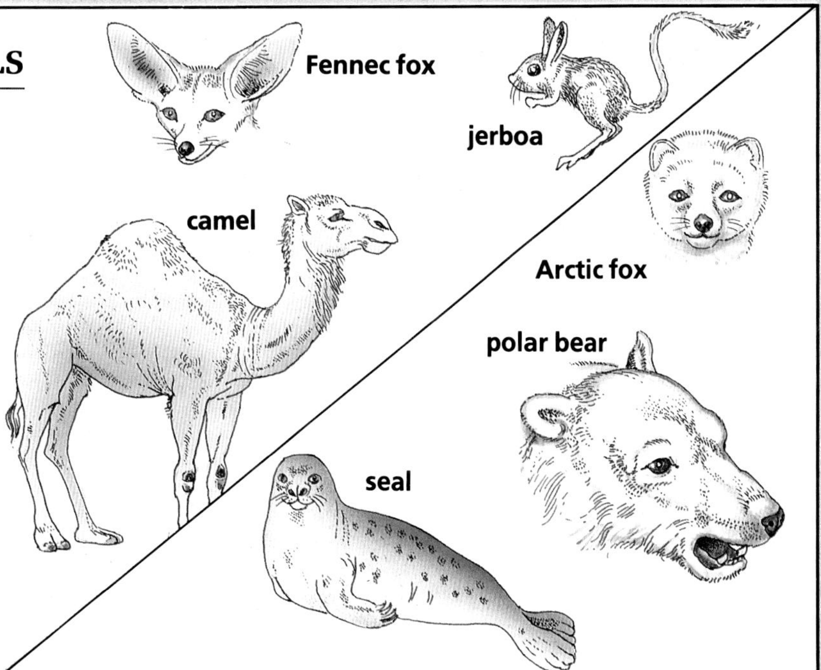

AIR PRESSURE

Air is all around you, but you cannot see it. Air is heavy and the weight of it pressing down on Earth and on you is called air pressure. At sea level more than one kilogram of air presses down on each square centimetre of your body. Why doesn't it squash you flat? It doesn't because the air inside your body is pressing out just as hard. Air pressure is greatest at sea level, but as you go higher up there is less air so the air pressure is lower. Air pressure is constantly changing – it is a good indicator of the type of weather that can be expected.

HIGHS AND LOWS

Lows are areas of low air pressure, the lowest pressure is in the centre. They bring cloudy, wet weather. Winds blow anti-clockwise. Highs are areas of high air pressure, the highest pressure is in the centre. They bring sunny weather. Winds blow clockwise. Isobars are lines on a weather map joining places with equal air pressure.

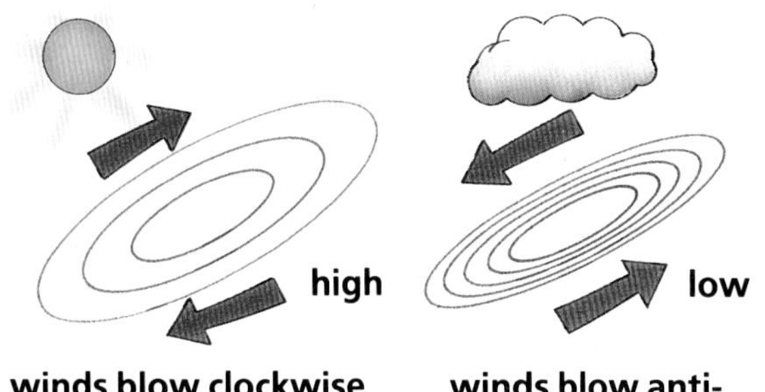

winds blow clockwise

winds blow anti-clockwise

Air pressure is measured with a barometer. If this "rises or goes up", air pressure is increasing. The weather is fair. If it "falls or drops", the air pressure is decreasing. It is unsettled and may rain.

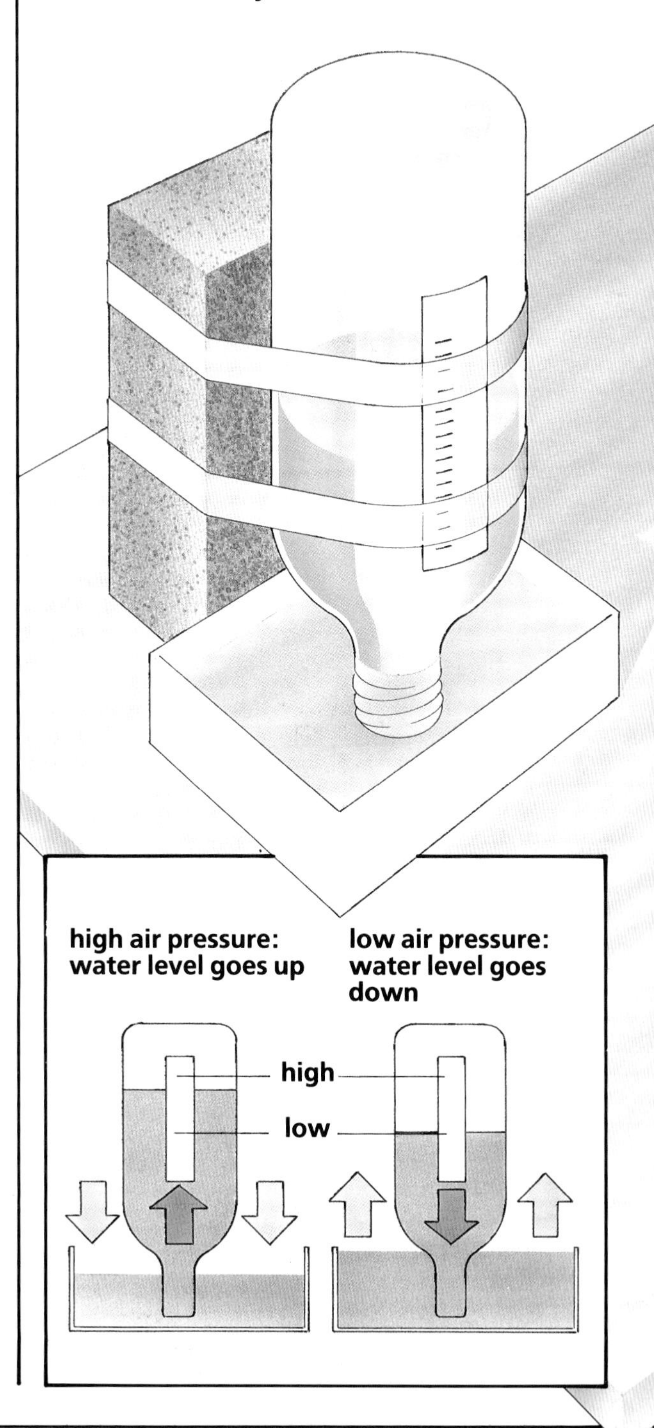

high air pressure: water level goes up

low air pressure: water level goes down

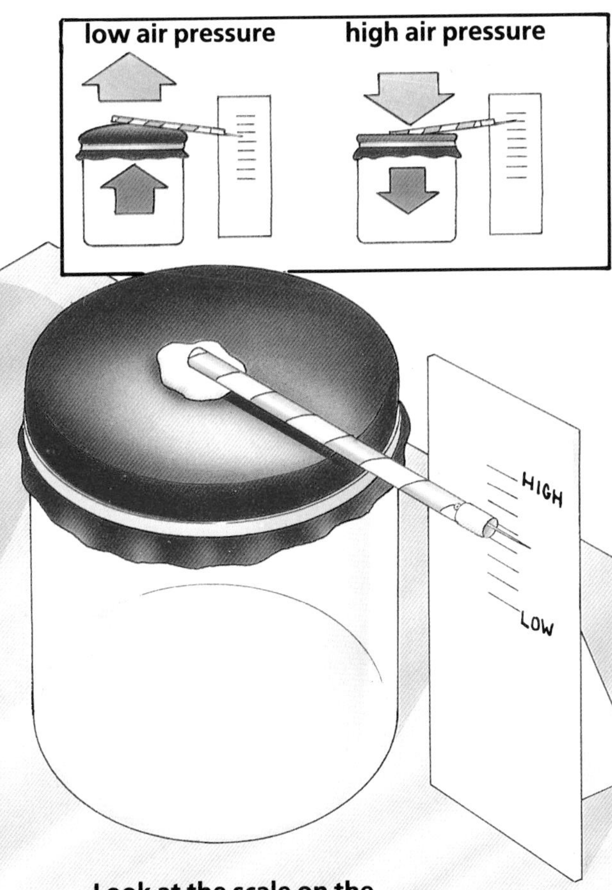

Look at the scale on the barometer at the same time each day. Write down whether the air pressure is rising, falling or steady and what the weather is like.

Date	Time	Pressure	Remarks
2 Aug	9am	Rising	cloudy warm
3 Aug	9am	High	warm sunny
4 Aug	8:30am	Steady	warm sunny humid
5 Aug	8:30am	Falling	Thunderstorm in afternoon

Put your readings in your notebook

MAKING A WATER BAROMETER

WHAT YOU NEED:

How to make it

1 Fill the container with 5–7 cm of water. Fill the bottle ¾ full of water.
2 Hold the card to the top of the bottle, turn it upside down, put it under the water, remove the card.
3 Stick some paper to the side of the bottle. Tape the bottle to the brick.
4 Mark the water level.

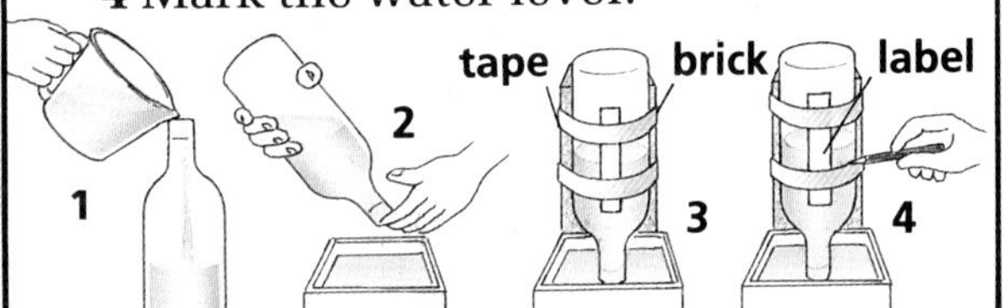

MAKING AN AIR BAROMETER

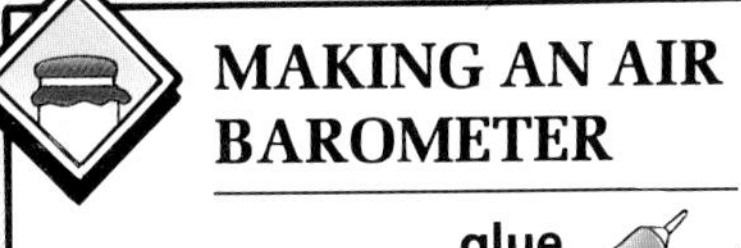

How to make it

1 Stretch the balloon over the jar.
2 Tape the pin to the end of the straw and glue to the balloon.
3 Fix your support to the card and write your scale.

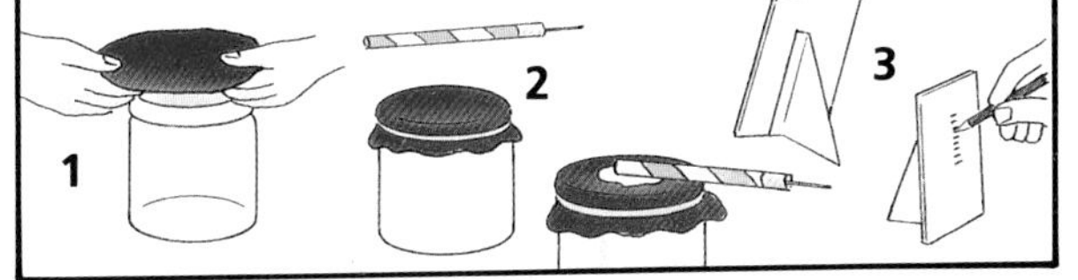

WHAT IS AIR?

Earth is surrounded by a layer of air known as the atmosphere. This is a mixture of gases, mainly nitrogen and oxygen, with small amounts of other gases like carbon dioxide and water vapour. Clouds that give rain and snow are formed from water vapour in the air.

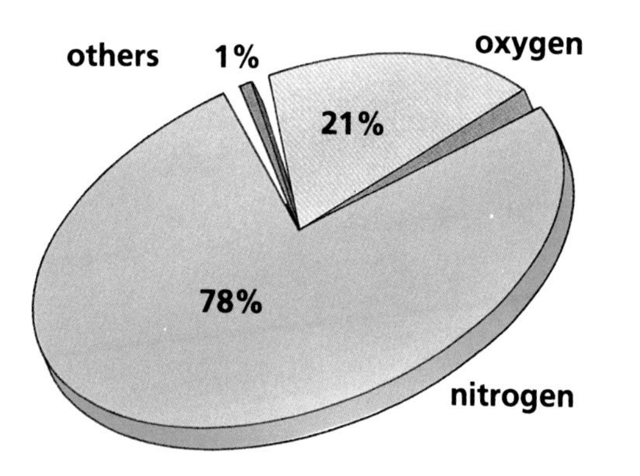

AIR POWER

Put a ruler on the edge of a table, spread a newspaper on top and hit the ruler with your fist. You will find it hard to lift the paper. Air pressure pushes down on the paper and stops the ruler lifting it. The paper has a large surface area so a lot of air is pressing down.

IS AIR HEAVY?

Blow up two balloons, making them the same size. Tie them to the ends of a coat hanger and hook this onto a cane. When they are balanced, burst one of them. The full balloon dips down because it contains air, so it is heavier than the burst balloon.

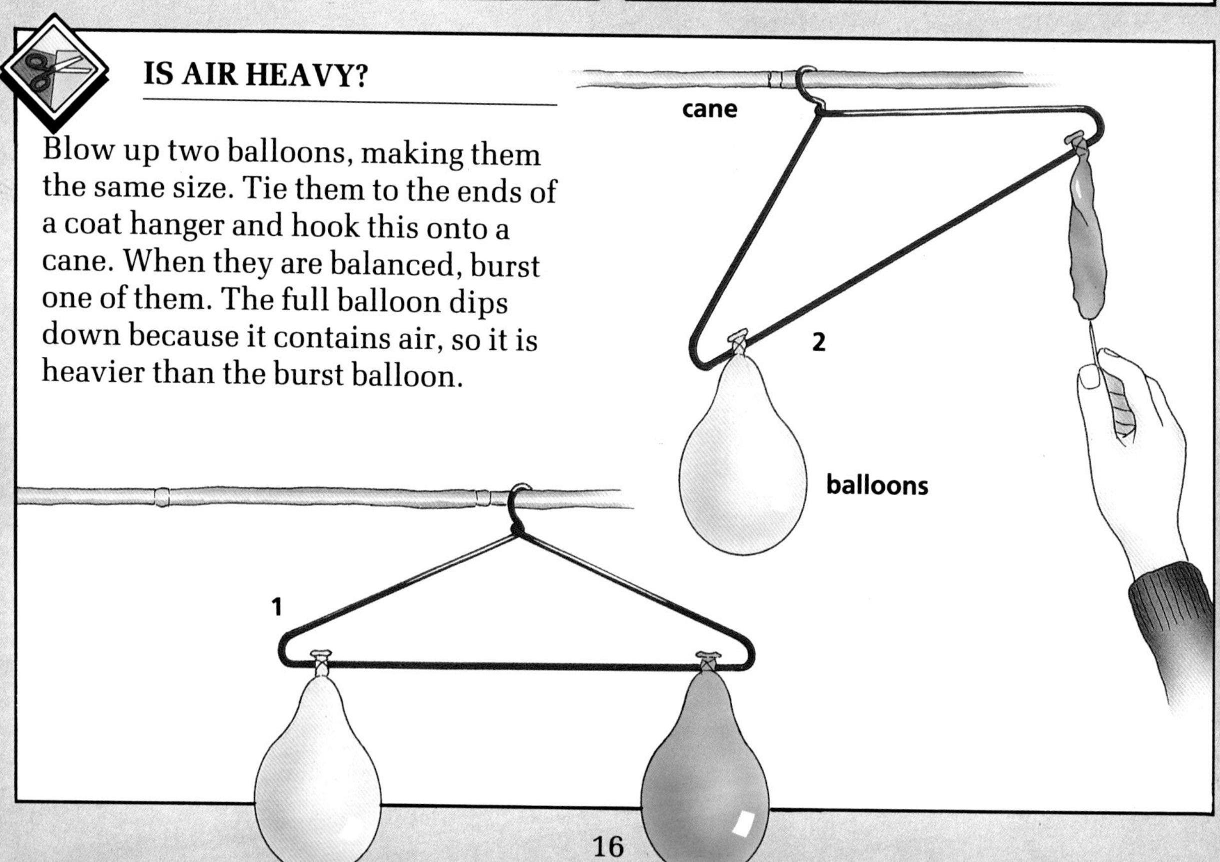

GLASS TRICK

Fill a bowl with water. Screw up a paper towel and wedge it in a glass. Turn the glass upside down and push it down into the water until the glass is submerged. Take the glass out and you'll see that the paper towel is dry. Air in the glass presses down on the water and stops it rising. Air can press in all directions, including upwards.

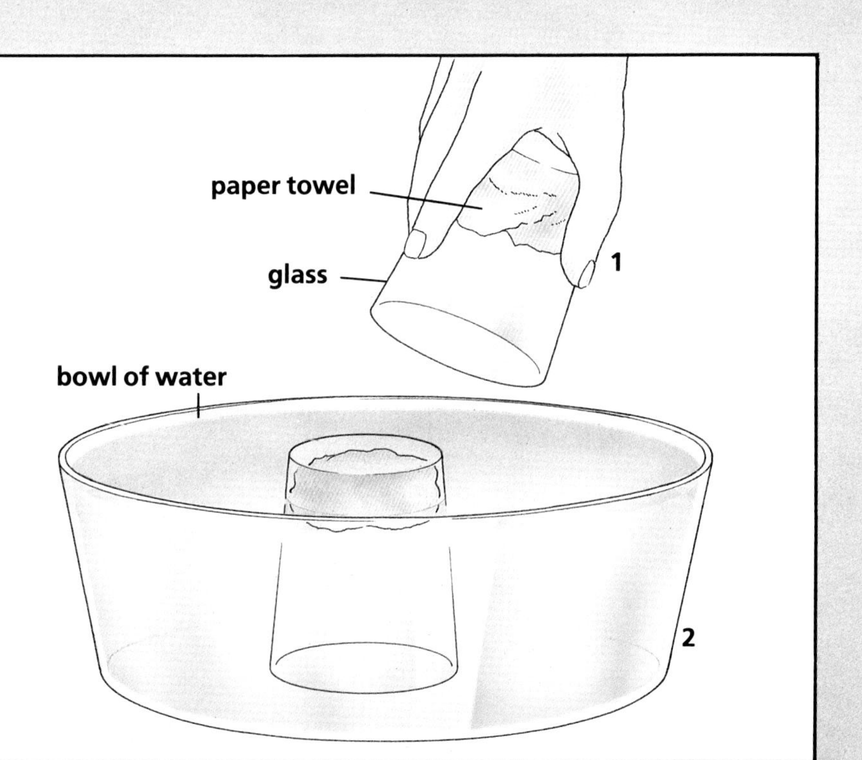

BAROMETERS

Barometers measure air pressure in units called millibars (mb). A barometer has a needle on a dial marked in millibars. This moves if the air pressure changes. If it rises, expect sunny, settled weather, if it falls it means clouds, rain and wind are due.

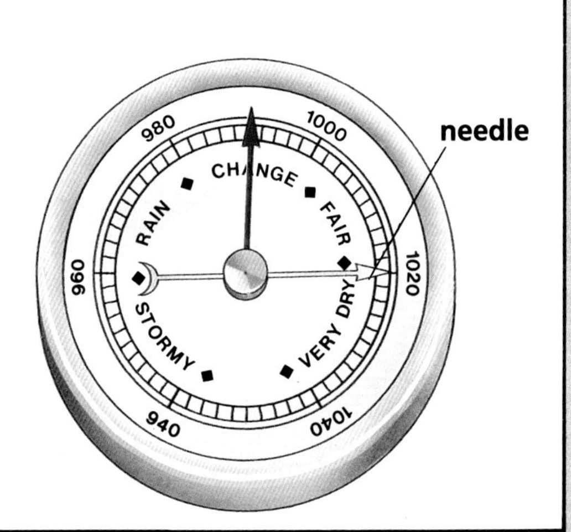

DID YOU KNOW?

Mountaineeers may have to use oxygen from cylinders if they are climbing very high mountains. As they climb there will be less and less air so there will not be enough oxygen for them to breathe.

Red cells in the blood carry oxygen from the lungs to the rest of the body. People who live at high altitudes (high above sea level) have more red cells so they can take up as much oxygen as possible. Athletes often train at high altitudes to build up more red cells.

Aeroplanes have specially pressurized cabins because they fly at a high altitude where there is not enough oxygen in the air for people to breathe.

WINDS

Wind is moving air. It is formed when the Sun heats the land and sea, and they heat the air above them. This hot air is lighter and rises, so it doesn't press so hard on Earth and a low pressure area forms. As air cools, it becomes heavier and sinks and a high pressure area forms. When the wind blows, two things are important. How strongly is it blowing (wind speed); and where is it blowing from (wind direction)?

GLOBAL WINDS

Winds do not blow in straight lines between the poles and the Equator. Earth spins on its axis like a top, which makes the winds swerve. In the northern hemisphere (top half) they swerve to the right, in the southern hemisphere (bottom half) they swerve to the left.

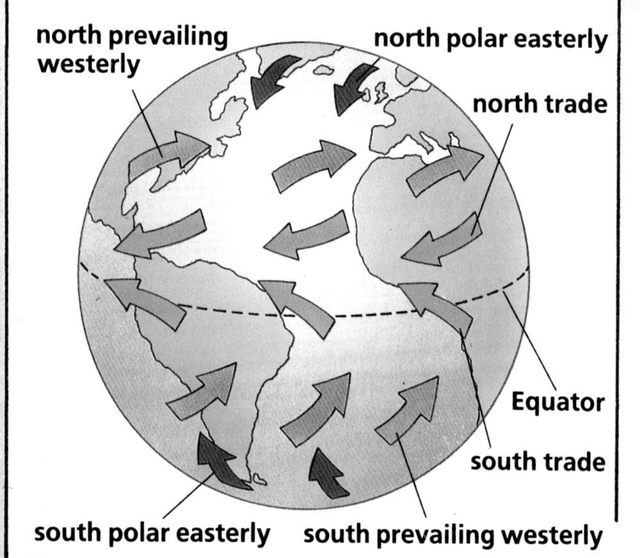

A weather vane, or wind vane, is used to show wind direction – the direction from which the wind is blowing. (A north wind, for example, blows from the north, not towards the north.)

WIND ROSE

You can fill in a wind rose to record wind direction. Using your vane at about the same time each day, work out the wind direction, find the correct compass bearing on the wind rose and shade a square in. Do this every day for a month and find where the wind blew from most often. This is called the prevailing wind and in Europe it is a south-west wind, that means it blows from the south-west.

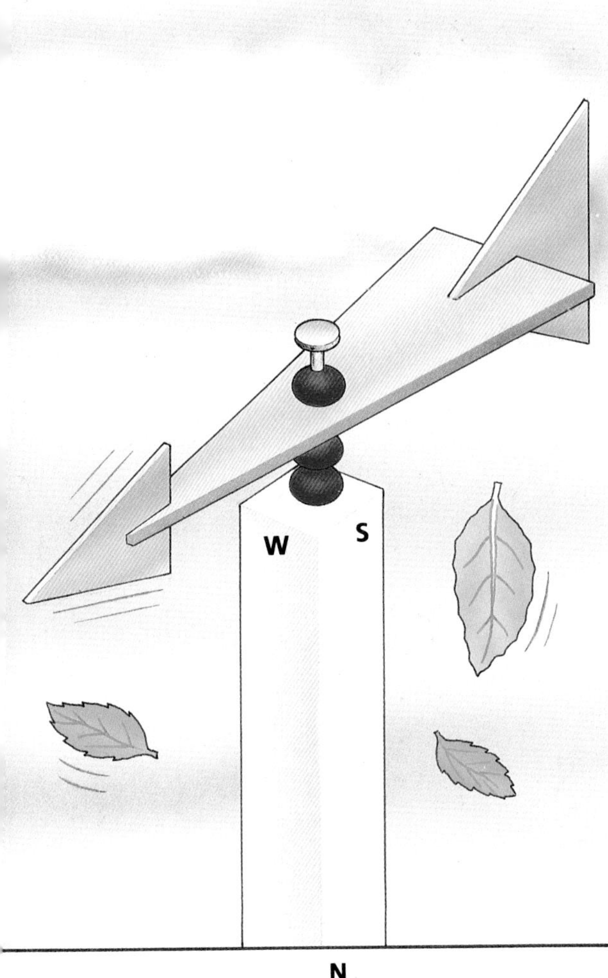

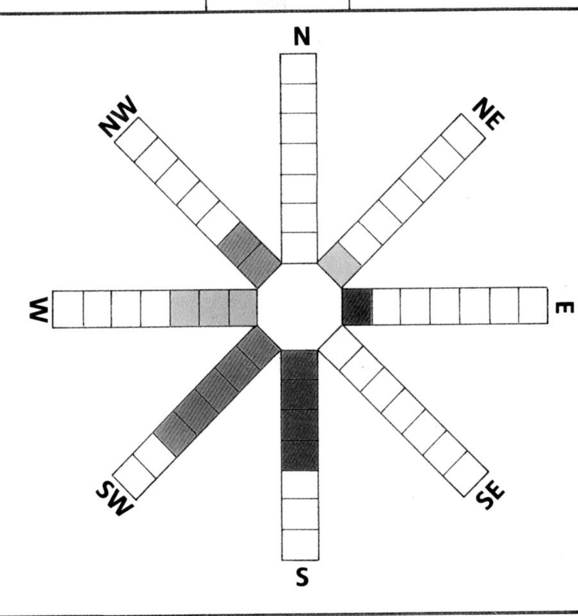

MAKING A WIND VANE

How to make it

1 Ask an adult to cut the wood into three shapes with the craft knife.

2 Find the centre of the middle section, mark it, then glue the front and back pieces in position.

3 Balance the vane on your finger, find the centre, mark it and carefully make a hole through it with the nail.

4 Using a compass, mark the points of the compass on top of the post.

5 Thread the beads and vane onto the nail, and ask an adult to hammer it into the top of the wooden post.

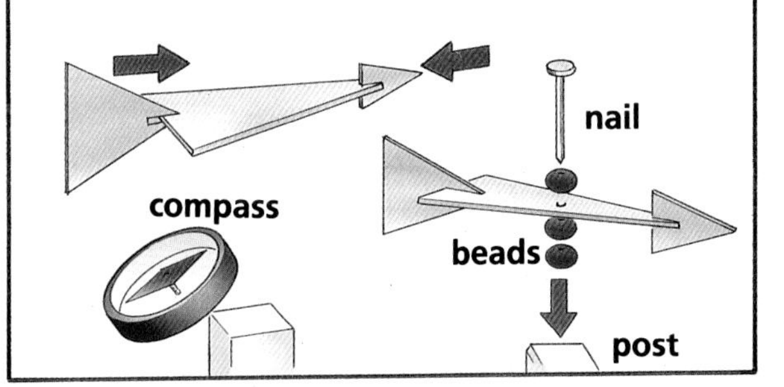

CUP ANEMOMETER

The instrument used to measure wind speed is called an anemometer. They are expensive to buy, but here are two that you can make.

WHAT YOU NEED:

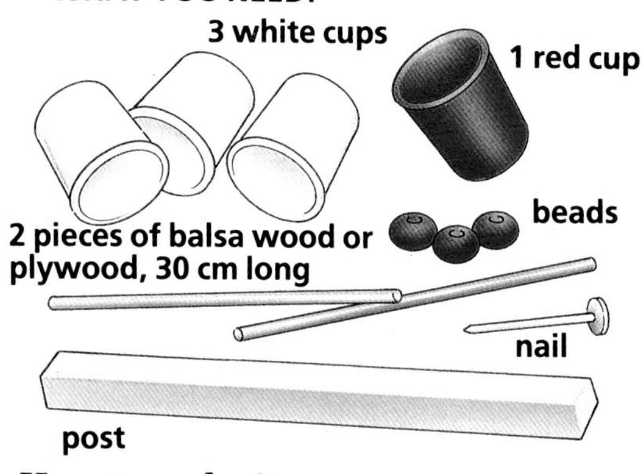

How to make it

1 Glue the two pieces of balsa or plywood together at right angles to each other. When they are stuck firmly push a nail through the centre (if you are using plywood, ask an adult to drill a hole).

2 Glue the base of the pots to the ends of the wooden arms, making sure the pots all face in the same direction.

3 Thread the beads and anemometer onto the nail and ask an adult to hammer it into the top of the post.

How it works

The faster the wind blows, the faster the pots go round. Follow the red pot, counting how many times the anemometer turns in a minute.

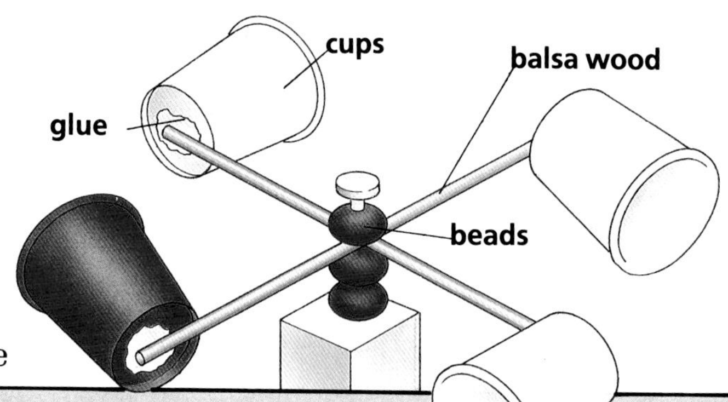

PROTRACTOR ANEMOMETER

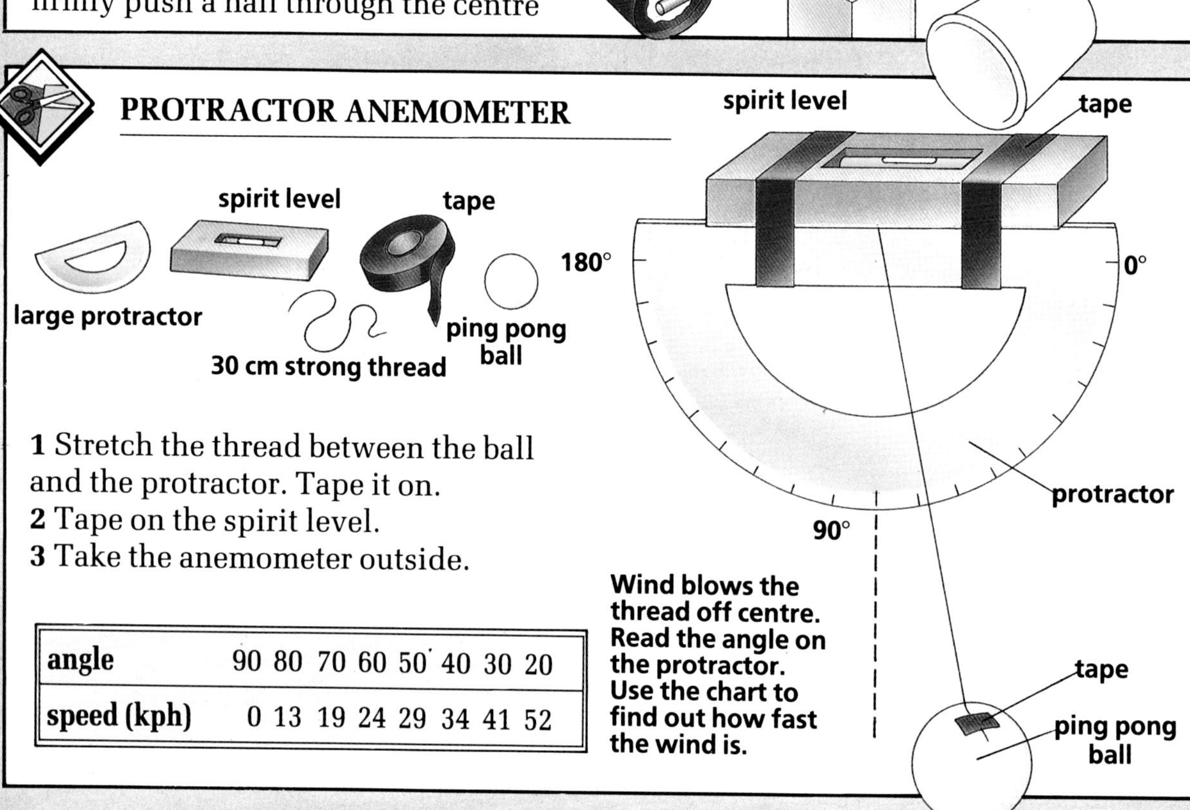

1 Stretch the thread between the ball and the protractor. Tape it on.

2 Tape on the spirit level.

3 Take the anemometer outside.

angle	90	80	70	60	50	40	30	20
speed (kph)	0	13	19	24	29	34	41	52

Wind blows the thread off centre. Read the angle on the protractor. Use the chart to find out how fast the wind is.

FRUITS AND SEEDS

Many fruits and seeds are spread or dispersed by the wind. Sycamore fruits have wings which slow down their fall as they spin away from the parent tree. Test this for yourself. Collect some sycamore fruits in autumn. Stand on a table or chair and drop one. Use a stop watch to time how long it takes to reach the floor. Take the wing off and see how long it takes to fall now.

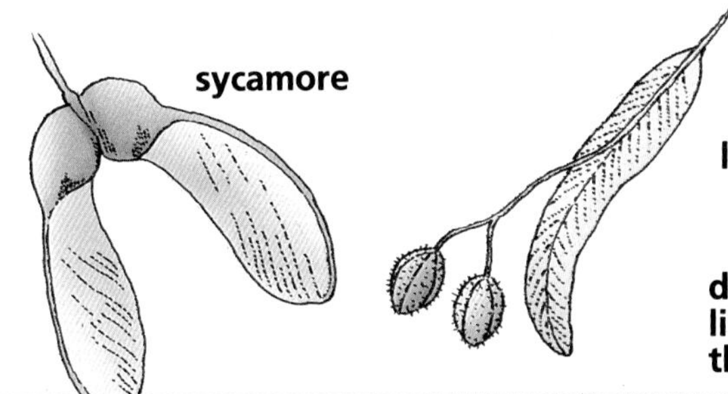

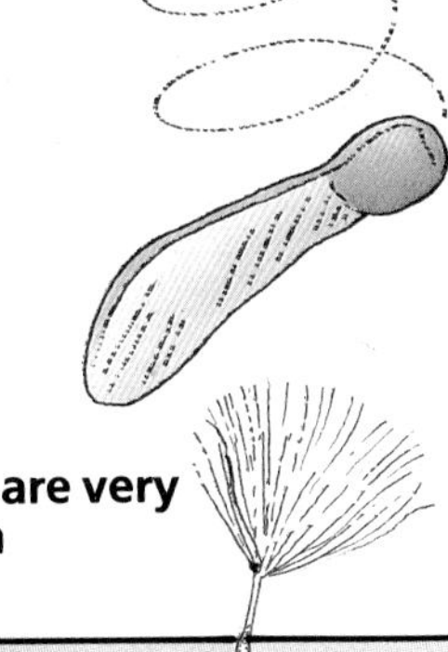

HURRICANES

Violent storms are called hurricanes in the Caribbean, typhoons in the China Seas, cyclones in the Indian Ocean and willy-willies off Australia. They spin as they travel, with winds up to 300 km per hour. Scientists give hurricanes names.

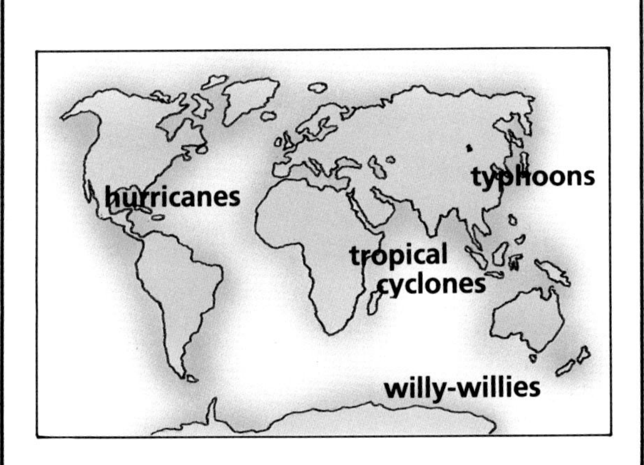

DID YOU KNOW?

The windiest place in the world is the George Fifth coast in Antarctica. Winds of 320 km per hour have been recorded there.

Jet streams are very strong winds blowing about 10 km above Earth. They can be thousands of kilometres long and no more than 500 km wide. There is a very strong jet stream over the North Atlantic. If aeroplanes are flying from America to Europe, in the stream, it can take about an hour off the flying time.

CLOUDS

Clouds are made of billions of tiny water droplets or ice crystals. The three main groups of clouds have Latin names: cirrus, meaning "curl of hair", cumulus, meaning "heap", and stratus, meaning "layer." They may have other names added to them, for example alto, meaning "high up", or nimbus, meaning "carrying rain". Each type of cloud carries a message about the weather.

HOW CLOUDS FORM

Air rises when it is warmed by the land or sea, when it is blown up and over hills, or when cold air flows under warm air, forcing it up.

Clouds form when warm air, carrying water vapour which has evaporated from the land and sea, rises. When this meets cold air the water vapour condenses on particles of dust, smoke or salt to form tiny water droplets.

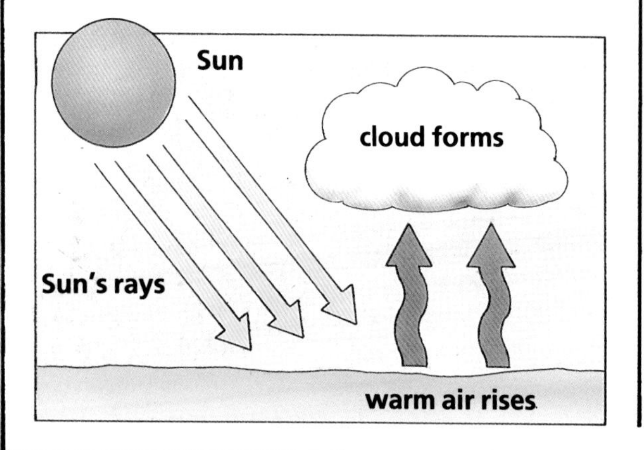

CIRRUS
Wispy streaks high in the sky, they indicate a change in the weather.

ALTOSTRATUS
Thin grey layers high in the sky; they can form rain or snow clouds.

CUMULUS
Puffs of fluffy cloud appear on sunny days.

Date	Time	Cloud type	Remarks
3 July	11am	none	sunny
	4pm	cumulus	sunny intervals
	8pm	cirrus	fine-warm
4 July	9am	stratus	overcast
	12am	altostratus	raining

CUMULONIMBUS
Huge towers in the sky; they bring thunderstorms with rain, hail or snow.

STRATUS
Grey sheets of low cloud that bring light rain. Fog is low level stratus cloud.

CLOUD IN A BOTTLE

WHAT YOU NEED:

How to make it

1 Fill the bottle with hot water, leave it for several minutes, then pour most of the water away.
2 Put an ice cube on top of the bottle.
3 Put the paper behind the bottle and watch what happens.

What happens

A cloud forms in the top half of the bottle. Warm, moist air at the bottom of the bottle, rises and meets cold air under the ice cube, so water vapour in the air condenses to form a cloud.

RAIN

Water covers over two-thirds of Earth's surface. It can exist as a gas, called water vapour, as a liquid, and as a solid, called ice. Raindrops form in clouds when billions of tiny water droplets move about and join to form larger droplets which become so heavy they fall as rain. When water falls as rain, hail or snow it is known as precipitation.

THE WATER CYCLE

The water on Earth is used over and over again as part of the water cycle.

The Sun heats oceans, lakes, plants and soil; water evaporates from them and rises into the air as water vapour (**1**). The water vapour cools and condenses to form water droplets in clouds (**2**). The water in clouds falls as rain or snow (**3**).

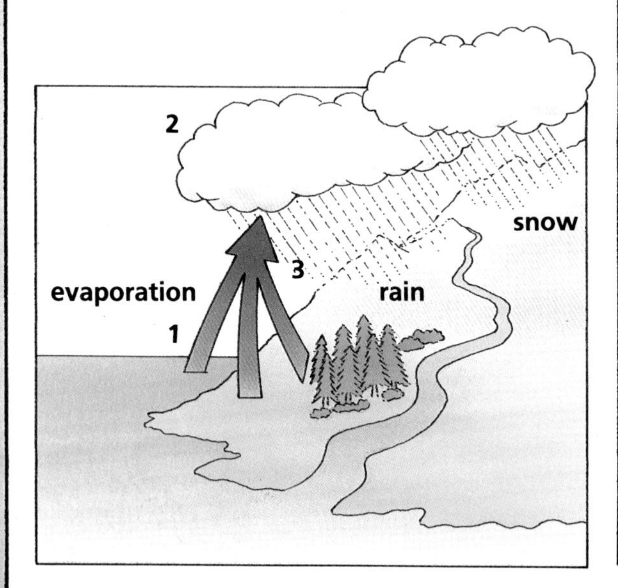

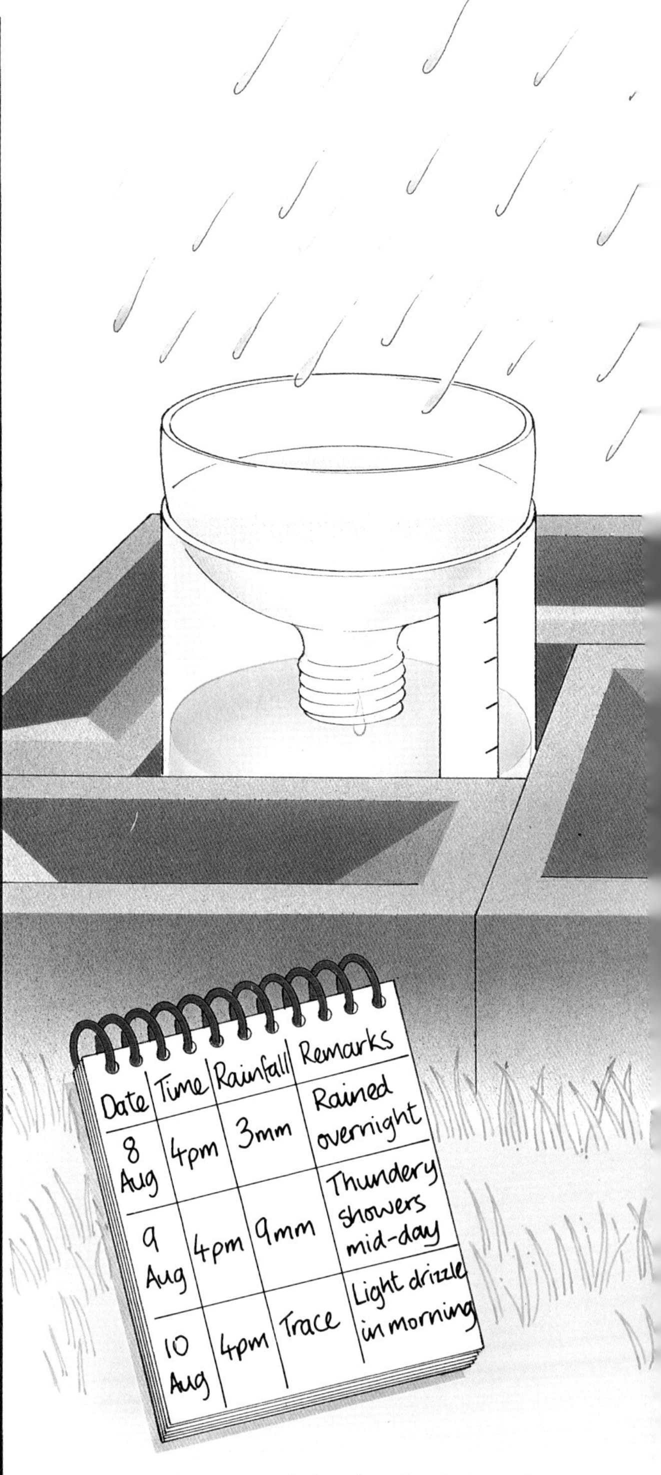

Every morning record the depth of the rain in millimetres. If the amount is too small to measure, it can be recorded as a "trace" in your notebook.

A rain gauge measures the depth of rain or snow which would cover the ground if none of it drained away or evaporated. It should be placed in the open, out of the wind, away from walls, fences and overhanging trees. To get more accurate readings with your rain gauge you can use a narrow bottle for measuring. Pour 1 cm of water into your narrow container. Mark off the level at 1 cm on your scale and divide into millimetres. Pour another centimetre in and mark your scale.

MAKING A RAIN GAUGE

WHAT YOU NEED:

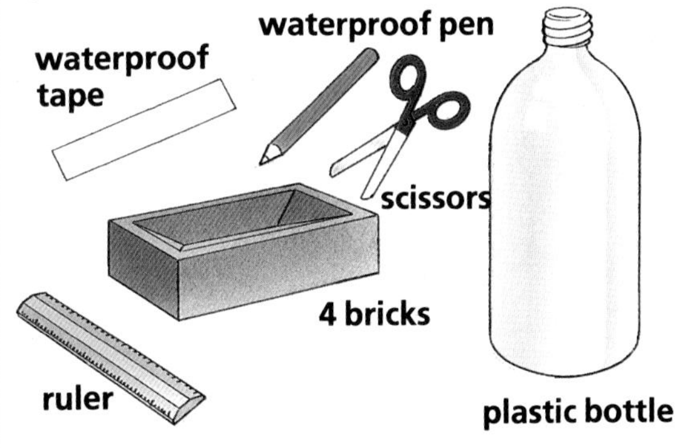

How to make it

1 Cut the top off the bottle.
2 With a ruler and pen, draw a scale in centimetres on waterproof tape and stick it up the side of the base.
3 Fit the upturned top into the base.
4 Put the gauge in the open supported by bricks. Record the rainfall each morning, empty the gauge and reset.

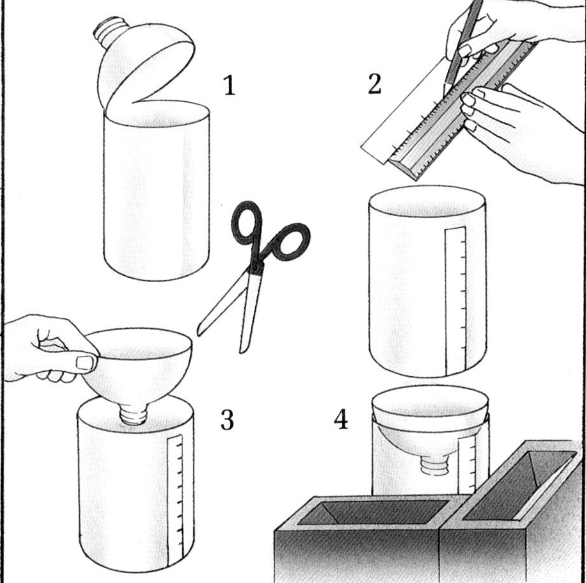

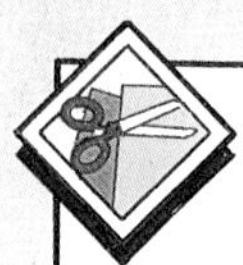

WATER FROM THE AIR

If you put a glass of water in a refrigerator for an hour, then take it out, you will see drops of water on the outside of the glass. The cold glass cools the air around it; cold air cannot hold as much water vapour as warm air, so water vapour changes into water droplets. When a vapour changes into a liquid it is known as "condensation".

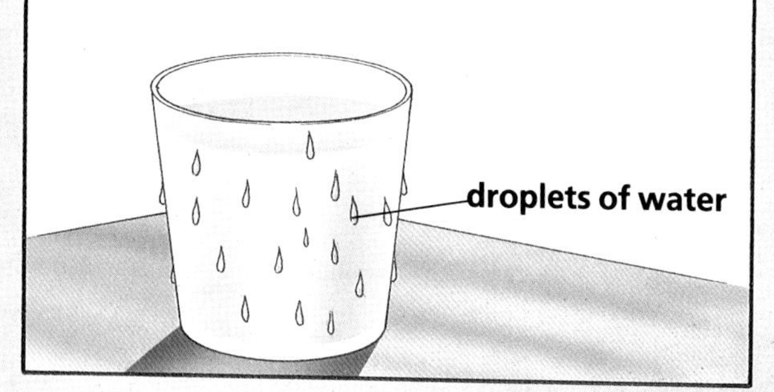

DISAPPEARING WATER

Put two saucers, each holding two spoonfuls of water, on a sunny windowsill. Shade one with a curtain. The water in the Sun dries more quickly. Heat from the Sun makes the liquid water evaporate – it changes into a vapour.

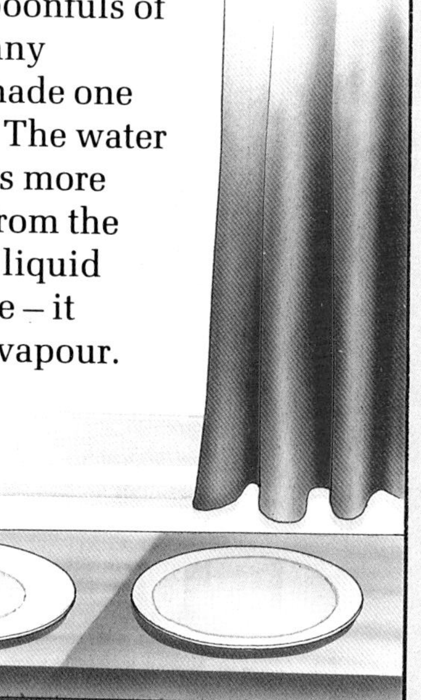

MAKING RAIN

Ask an adult to help you use a kettle. Put a large metal spoon in the freezer, then boil the kettle and place a saucer under the spout. Put on an oven mitt, then hold the spoon in the vapour from the kettle. Watch the rain falling. Water vapour from the kettle hits the cold spoon and condenses to form water droplets.

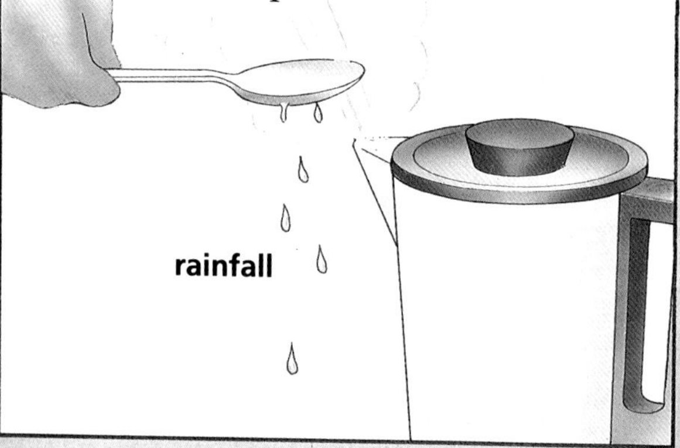

DID YOU KNOW?

The wettest place in the world is Mount Wai-'ale-'ale in Hawaii; the annual rainfall is 11,684 millimetres and it rains for about 335 days each year.

The driest place is Arica in Chile, with an annual rainfall of about 0.8 millimetres.

Raindrops are not tear-shaped, but are like flat-bottomed circles; they are better described as "bun shaped". They range in size from 1.5 millimetres to 5 millimetres in diameter (about the size of a pea). Droplets of drizzle have a diameter of less than 0.5 millimetres.

WATER SURVIVAL KIT

Try this simple way of making clean water from muddy or salt water.

WHAT YOU NEED:

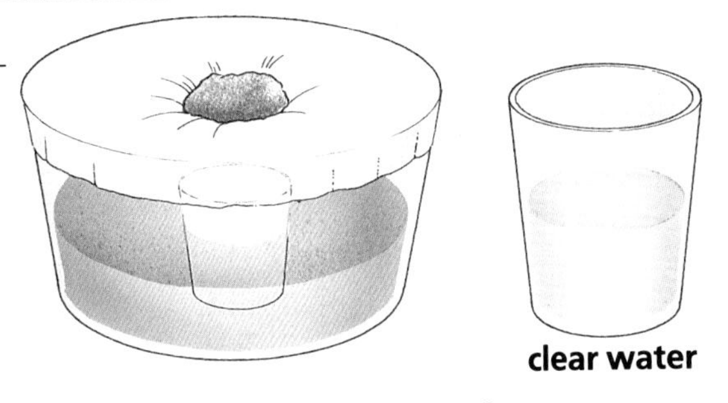

How to make it

Fill your bowl a third of the way up with muddy or salt water. Put the glass upright in the middle. Pull plastic wrap over the bowl. Put a stone in the middle of the plastic to make it sag, but it must not touch the glass. Put it in a sunny place.

What happens

At the end of the day the glass will contain clear water. The muddy water evaporates. Water vapour rises and touches the cooler plastic wrap. It condenses to form droplets that roll down to the centre of the plastic and drip into the glass. This way of purifying water is called distillation.

HUMIDITY

Humidity is the amount of water vapour or moisture in the air. Warm air holds more water vapour than cold air and the more water vapour there is the more humid it is. When humidity is high and it is warm, we feel sticky. To measure humidity you can make a hygrometer.

How to make it

1 Wrap two thermometers in cotton wool and secure with rubber bands.
2 Dip the cotton wool of one thermometer in water and keep it wet by dipping the end in a pot of water.
3 Hang the thermometers side by side in the shade or attach to the inside of a Stevenson screen.
4 Take the temperature of both thermometers every day.

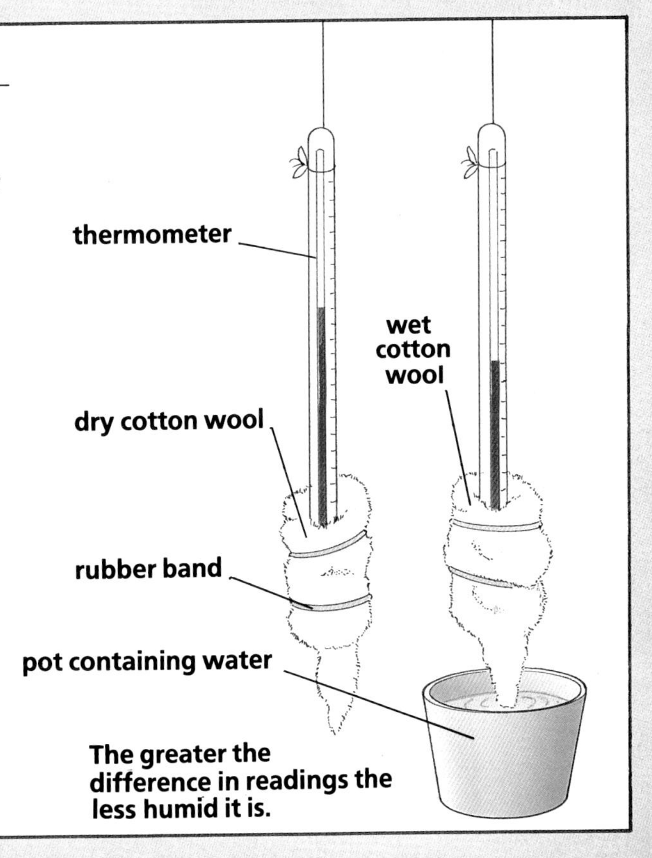

The greater the difference in readings the less humid it is.

RAINBOWS

Rainbows occur when the Sun is shining and it is raining at the same time, and sunlight passes through the raindrops. Sunlight is usually seen as white light but it is really made up of red, orange, yellow, green, blue, indigo and violet – the colours of the spectrum. When a ray of sunlight enters a raindrop it is bent or refracted and split up into all its colours. These bounce off the back of the raindrop and separate out again into all its colours.

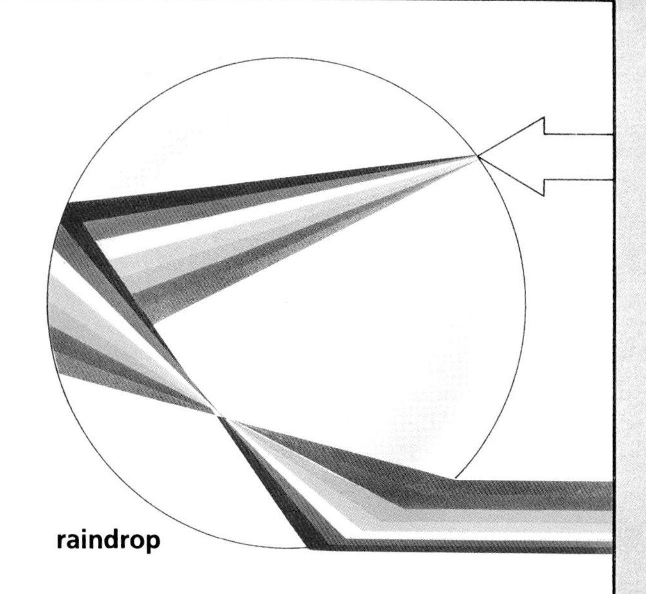

MAKE AN INDOOR RAINBOW

Put a glass of water on a sunny windowsill and a sheet of white paper below it, on a table. Move the paper around. Light passes through the water and is split into its different colours, forming a rainbow on the paper.

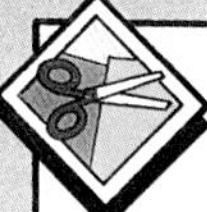

MAKE A GARDEN RAINBOW

Stand with your back to the Sun when it is low in the sky, turn on a garden hose and hold it up. Face something dark, like a bush, and look into the spray. Sunlight will enter the water drops and split the light to form a rainbow.

Hold the hose up, to get a fine spray

COLOUR SPINNERS

WHAT YOU NEED:

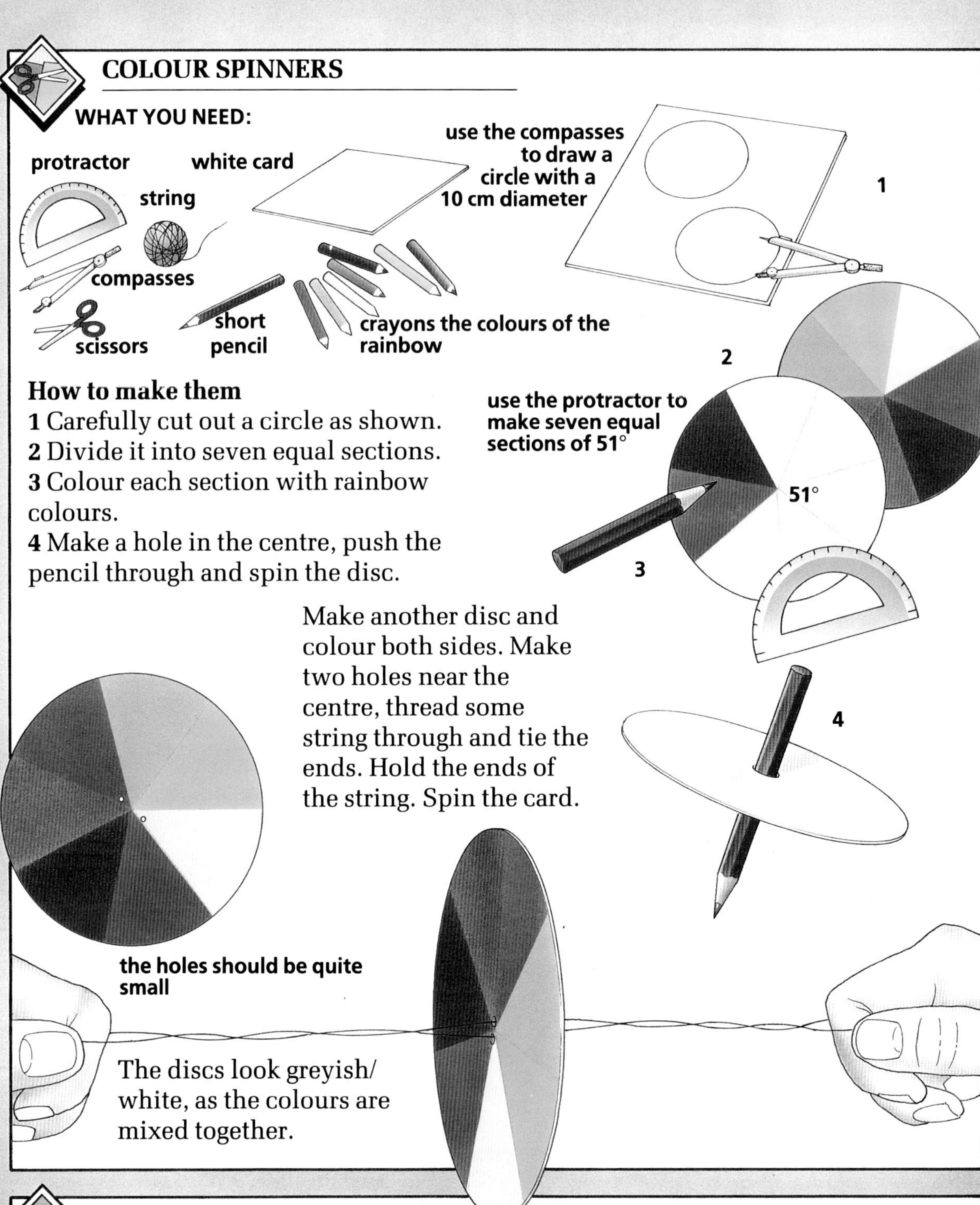

How to make them

1 Carefully cut out a circle as shown.
2 Divide it into seven equal sections.
3 Colour each section with rainbow colours.
4 Make a hole in the centre, push the pencil through and spin the disc.

Make another disc and colour both sides. Make two holes near the centre, thread some string through and tie the ends. Hold the ends of the string. Spin the card.

The discs look greyish/ white, as the colours are mixed together.

DID YOU KNOW?

Most rainbows last a few minutes but one in Wales lasted three hours.

From aeroplanes, rainbows appear as circles – red on the outside and violet in the middle. The circle appears between a cloud and the ground.

THUNDER AND LIGHTNING

Thunderstorms occur when large amounts of warm, moist air rise very quickly, then cool. Cumulonimbus clouds form. Inside a thundercloud fast-moving air causes the build up of positive (+) charges of electricity at the top of the cloud and negative (–) ones at the bottom. The ground is charged with positive electricity. Eventually a spark of electricity jumps the gap between the negative and positive charges in or between the clouds or from a cloud to the ground. We see this as lightning. The air becomes very hot and expands violently to produce thunder.

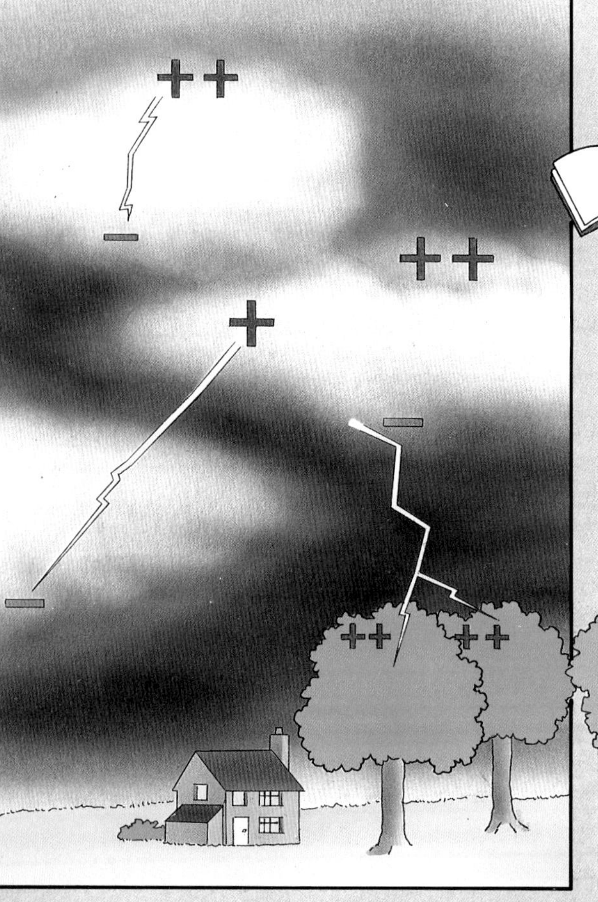

MAKE YOUR OWN LIGHTNING

WHAT YOU NEED:

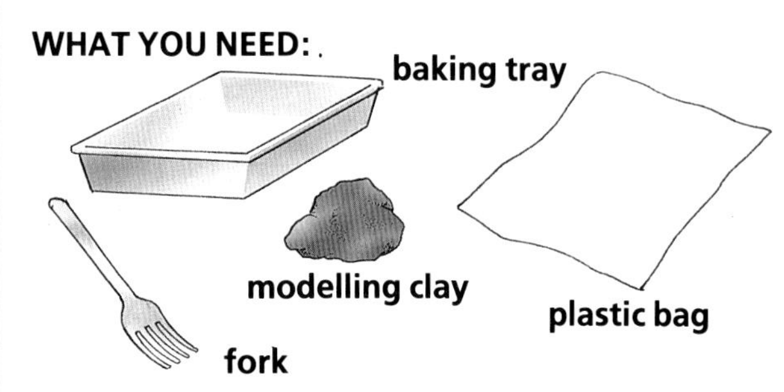

How to make it

1 Press the modelling clay into the middle of the tray. Put the tray on the plastic bag and, holding the clay, rub the tray round on the bag.

2 Pick up the tray using the clay.

3 Bring the fork towards a corner of the tray. A spark will jump from the tray to the fork. Static electricity has built up. When it is released it forms a spark.

HOW SAFE ARE YOU?

Lightning takes the shortest route to the ground so trees are often hit. **NEVER** shelter under trees in a storm. Move away from high ground. You are safe in cars, houses and aeroplanes as the electricity runs round the outside of them.

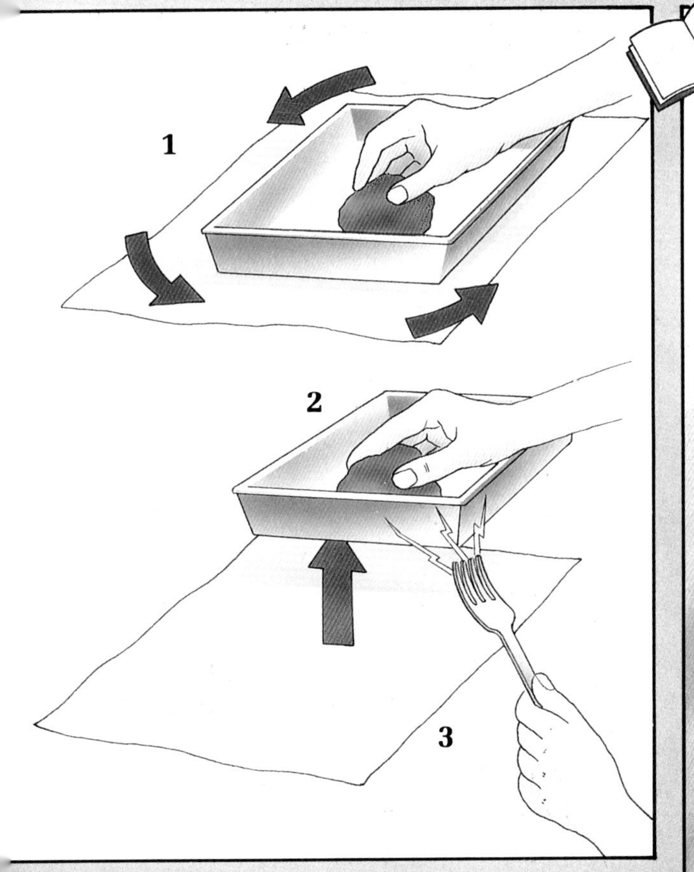

LIGHTNING CONDUCTORS

These are used to protect high buildings from damage by lightning. They are made of a strip of metal, such as copper, which runs from the highest point of the building, down the side to the ground. They carry electricity safely to the ground.

HOW FAR AWAY?

Thunder and lightning occur at the same time. Light moves faster than sound so lightning is seen before thunder is heard. To find out how far away the storm is count the seconds between the lightning and thunder and divide by 2½. Ten seconds means the storm is 4 km away.

DID YOU KNOW?

The Empire State Building in New York does not have a lightning conductor. It can be struck by lightning 500 times a year, and has been hit 48 times in one day.

Each year there are 16 million thunderstorms in the world, with about 1,800 raging at this moment.

Bogor in Java has had as many as 322 storms in one year.

SNOW AND ICE

Snowflakes are made up of ice crystals. They form in clouds that are colder than 0°C, the freezing temperature of water. When water freezes onto tiny ice particles in the air, it forms bigger ice crystals. The crystals fall through the cloud, bump into other crystals and become snowflakes. Falling snow melts if it meets warmer air, becoming rain.

SNOWFLAKE SHAPES

Most snowflakes have six sides and of the millions that have fallen to Earth nobody has ever found two exactly the same. The shape of the flake depends on the air temperature. In very cold air they are rod or needle shaped, but in warmer air pretty star-shaped or plate-like flakes form. Look at them through a magnifying glass.

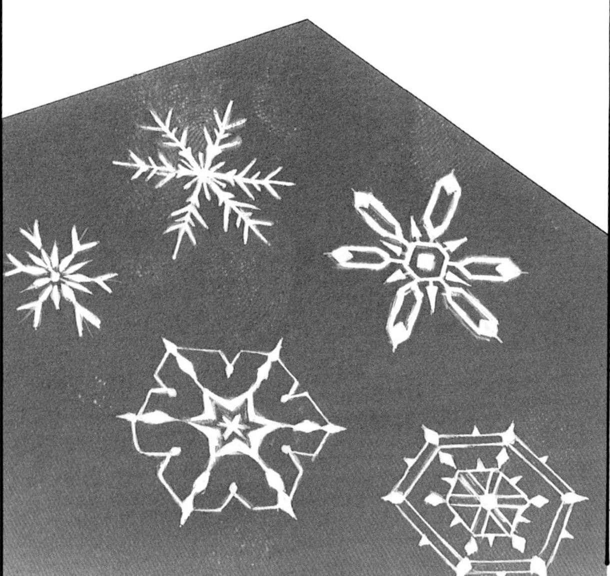

INVESTIGATING SNOW

Measure the depth of snow with a ruler in the open, not where the snow has built up into a drift or been blown away by the wind. If a lot of snow is falling, ask if you can go out to measure it every hour.

Fill a straight-sided container with snow. Don't squash it down. Measure the snow, then take it indoors and let it melt. When all the snow has melted, measure the depth of water produced. You should find that from every 10 centimetres of snow you will get about one centimetre of water. This is because air is trapped between the flakes and is released when the snow melts.

Scooping snow up into a container.

FLOATING ICE

As water freezes into ice, it expands and becomes lighter. This is why icebergs float. Ice takes up about a ninth more space than it did as water. So only about a ninth of an iceberg shows above the water, with nine times as much below.

When the ice cube melts, the glass will not overflow.

WHAT IS HAIL?

Hailstones are hard lumps of ice formed when air currents sweep ice crystals up and down in cumulonimbus (storm) clouds. Water freezes onto ice crystals in layers, like the skin of an onion. When they get quite heavy, they fall as hailstones. You can cut one in half to see the layers. Count these to see how many times the hailstone was tossed up and down in the cloud.

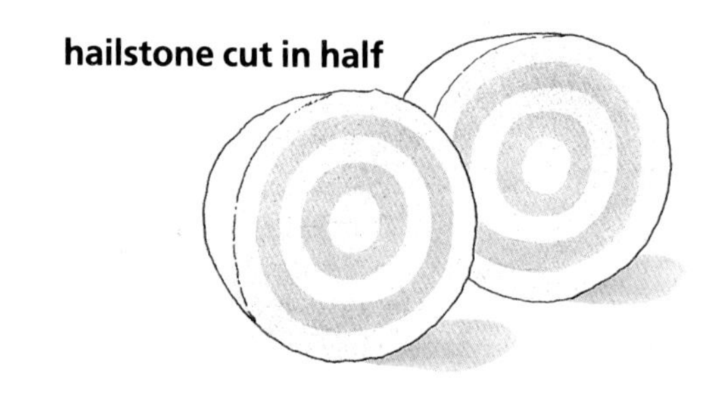

hailstone cut in half

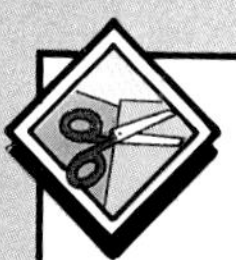

EXPANDING ICE

When the water inside water pipes freezes, the pipes often burst. You may not realize this until the ice melts and the water pours out. This happens because water expands when it freezes and so it becomes too big for the pipes. You can investigate this with a plastic container like an ice-cream tub. Fill it right to the top with water, replace the lid and put it in a freezer for a few hours. When you take it out, the lid has risen because the water inside it has frozen and expanded.

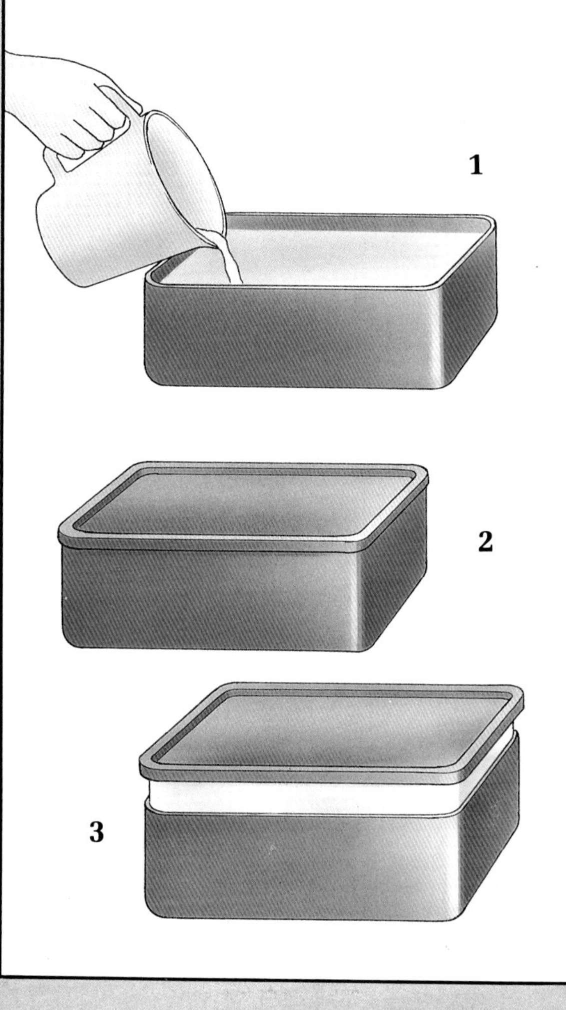

WINTER COATS

In winter some Arctic animals grow white coats which reduce heat loss and provide camouflage.

MAKING TRACKS

Looking for animal tracks in the snow can be great fun. See if you can identify the tracks. Here are some that you might come across.

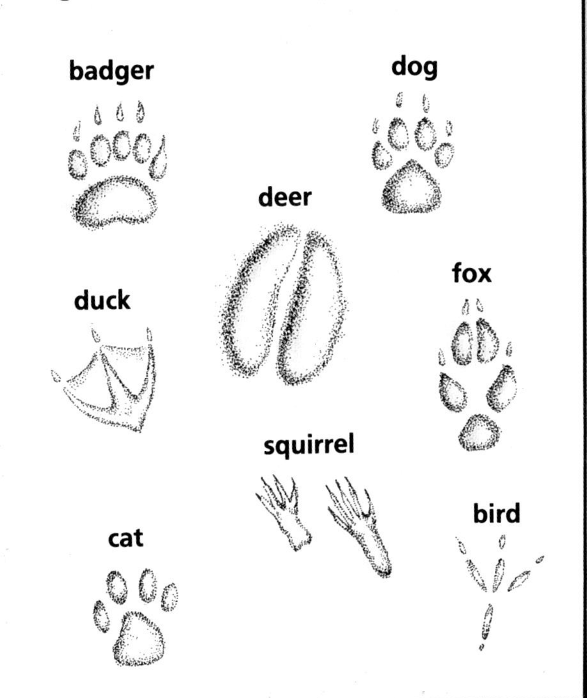

DID YOU KNOW?

It has been calculated that in a snowstorm 50 quadrillion snowflakes fall – that is 5 followed by 16 zeros (noughts)!

Canada, about half of the United States and Northern Europe have more snow each year than the North or South Pole.

An American nicknamed "Snowflake" Bentley spent 50 years studying snowflakes. He looked at thousands but never found two exactly the same.

Most hailstones are about the size of a pea, but some as big as melons have fallen. The largest hailstone has been recorded in Bangladesh weighing 1.02 kilograms – about the weight of a bag of sugar!

WEATHER FORECASTING

The study of the atmosphere and weather is called meteorology, and the scientists who study them are called meteorologists. They gather information about the weather from many different sources and piece it together to predict what the weather will be like. All the information is sent around the world.

COLLECTING INFORMATION

About 10,000 weather stations all over the world collect information on the weather. Along with readings from ships, planes, satellites and balloons, this information goes to weather centres, where it is plotted on a map called a synoptic chart. The information is fed into a computer and forecasts for the next few days are made.

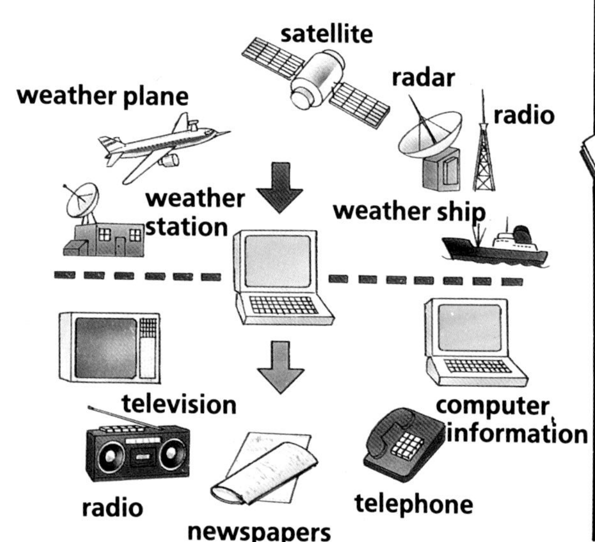

WARM FRONT
An area of rain followed by warmer weather.

HIGHS AND LOWS
An area of low air pressure (depression) means wet weather and strong winds. An area of high air pressure (anticyclone) brings dry weather.

WHAT IS RECORDED?

Meterologists record details of the following.

WHERE WEATHER HAPPENS

Earth is surrounded by a layer of gases which make up the atmosphere. This is divided into layers. The highest layer is the stratosphere. The lowest layer, the troposphere, is about 16 kilometres deep and it is where weather happens.

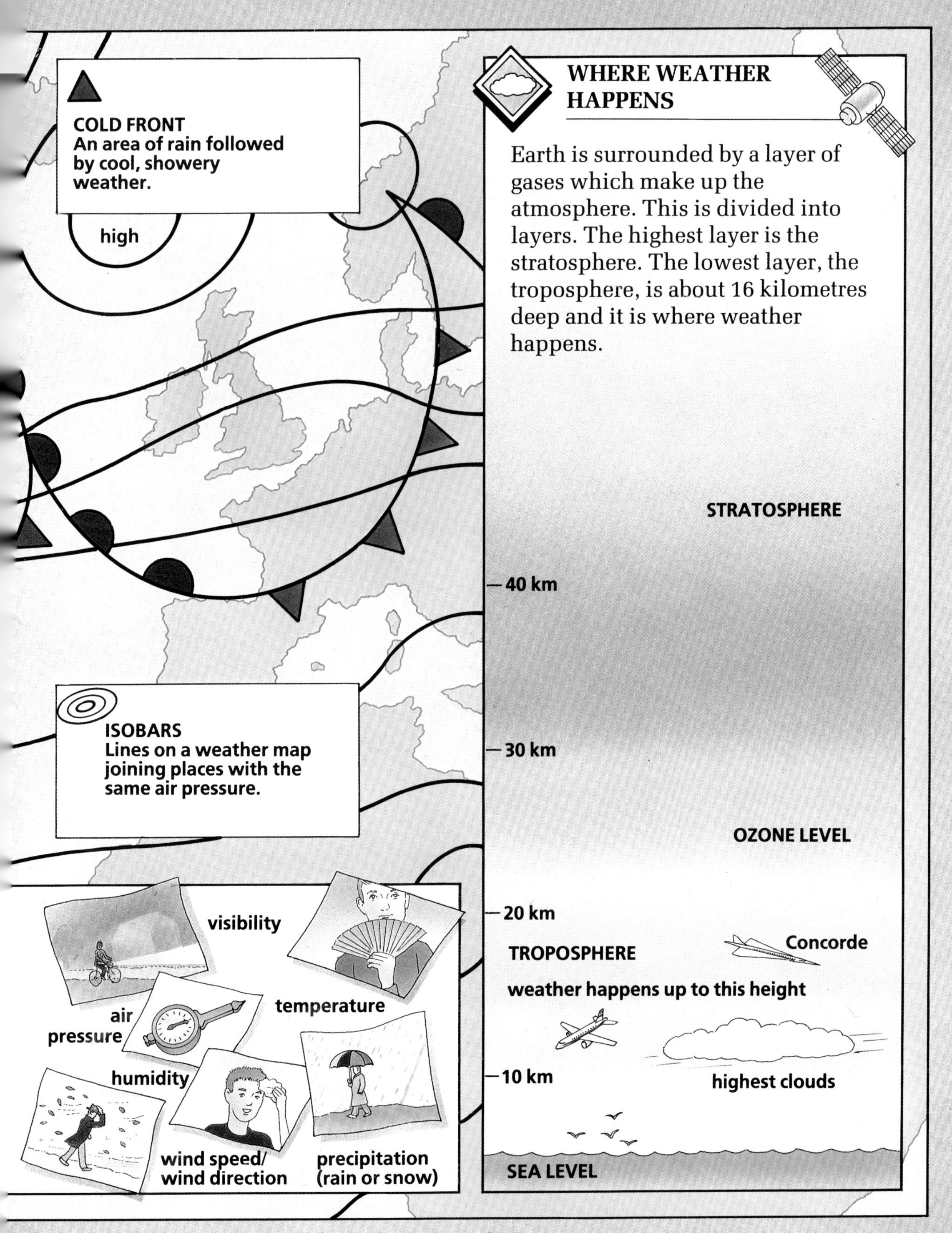

WEATHER SYMBOLS

Several different styles of weather map appear on the television and in newspapers and many of them use symbols. Here are a few you may recognize.

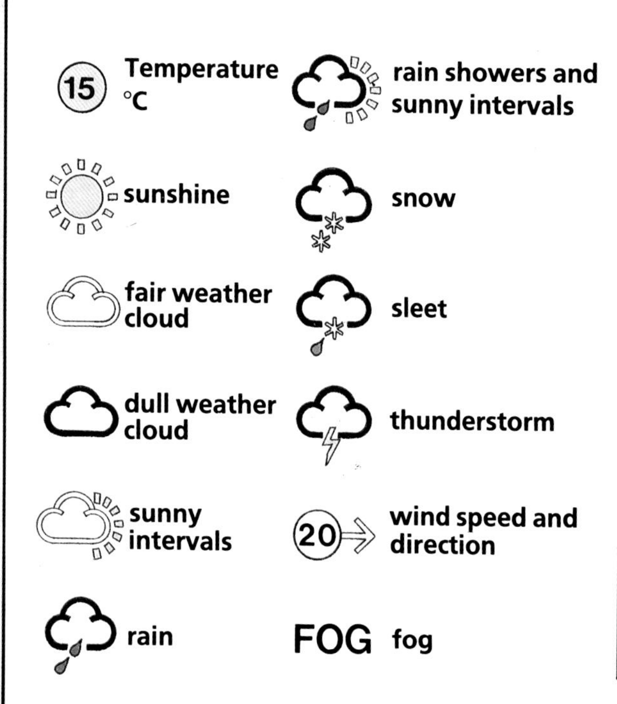

YOUR OWN WEATHER SHORTHAND

If you want to record the weather by using symbols as well as writing it in your log book, here is one way to try. Do this for a month, then count the days when the Sun shone or it rained.

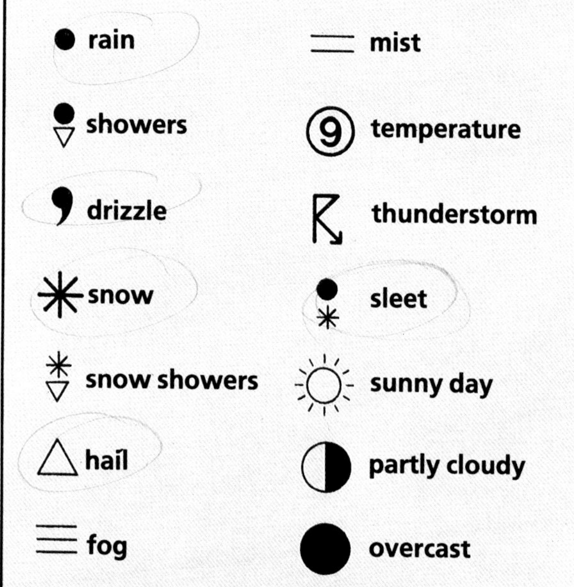

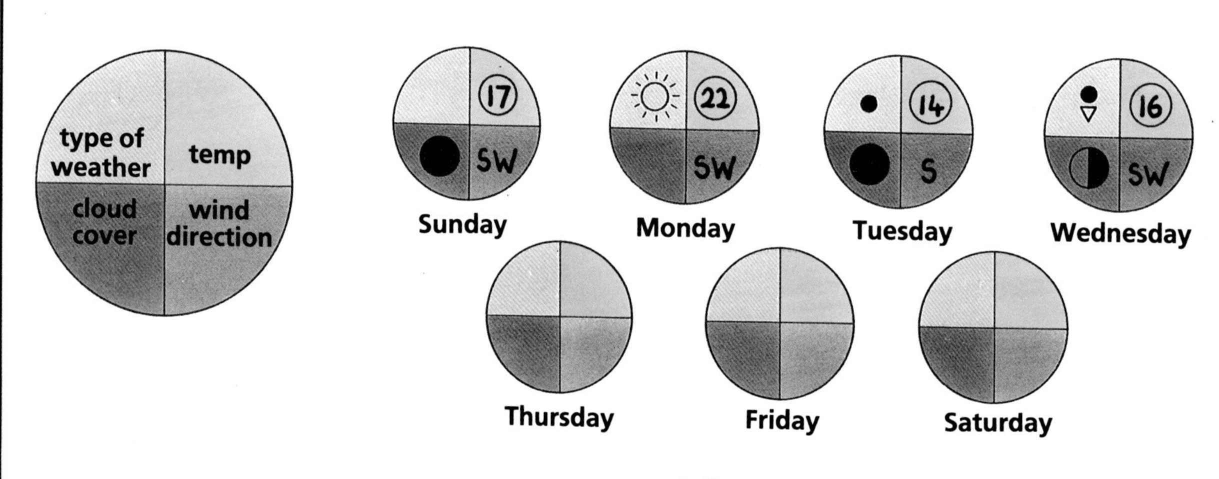

daily record chart

DID THEY GET IT RIGHT?

Weather forecasts are said to be right nine times out of ten, but why not check this out. From the newspaper, TV or radio, put down the forecast for that day. At the end of the day, or next morning, use your own readings, and the newspaper's report, to fill in actually what happened.

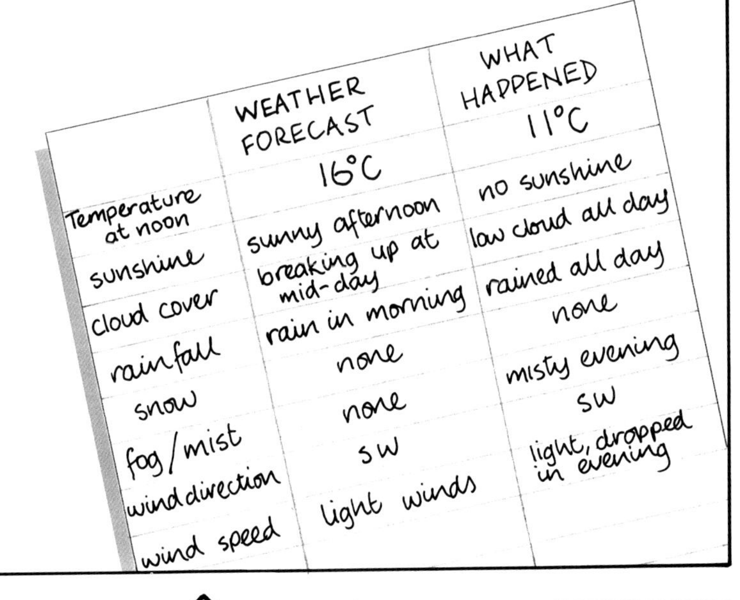

	WEATHER FORECAST	WHAT HAPPENED
Temperature at noon	16°C	11°C
sunshine	sunny afternoon	no sunshine
cloud cover	breaking up at mid-day	low cloud all day
rainfall	rain in morning	rained all day
snow	none	none
fog / mist	none	misty evening
wind direction	SW	SW
wind speed	light winds	light, dropped in evening

DID YOU KNOW?

In North America people say 2 February is Groundhog Day. If one wakes from its winter sleep, sees its own shadow (meaning it is sunny but cold) and goes back to sleep, winter will last another 6 weeks.

German people used to keep frogs as barometers – they croaked when the pressure dropped.

Animals and plants are said to give clues that the weather is changing. Dogs and farm animals are often restless before a storm. Some people say that cows lie down before rain, ants move quicker if it is hotter, pine cones close if rain is coming. Check these stories and see if they are true.

YOUR WEATHER SCRAPBOOK

If there is a heavy snowfall or strong gale in your area, cut out any article about it in the newspaper and put it in a scrapbook with any photographs and a weather map for that day. Don't forget to enter the date.

CHANGING CLIMATES

Over millions of years Earth's climate has changed very slowly, with cold periods (the Ice Ages) followed by warmer periods. At present the temperature is just right for life. In recent years, however, many scientists have found evidence that the world's climate is changing, and that this is because of human actions.

THE GREENHOUSE EFFECT

1 Carbon dioxide, produced when rain forests, coal and oil are burned, rises and forms a layer round Earth.
2 The Sun's rays hit Earth's surface.
3 Heat reflected back into the atmosphere meets the layer of carbon dioxide, trapping some of the heat which is reflected back to Earth. The world's temperature could rise by up to 4°C over 150 years.

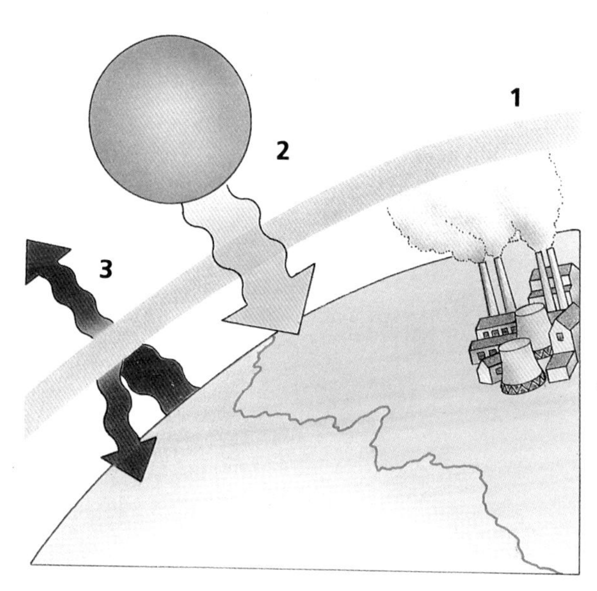

THE OZONE LAYER

Ozone is a gas like oxygen. It forms a layer 10 to 35 kilometres above Earth, filtering out much of the ultra-violet radiation from the Sun which would otherwise be harmful to plants and animals. The amount of ozone is falling. In 1985 scientists found holes in the ozone layer over Antarctica.

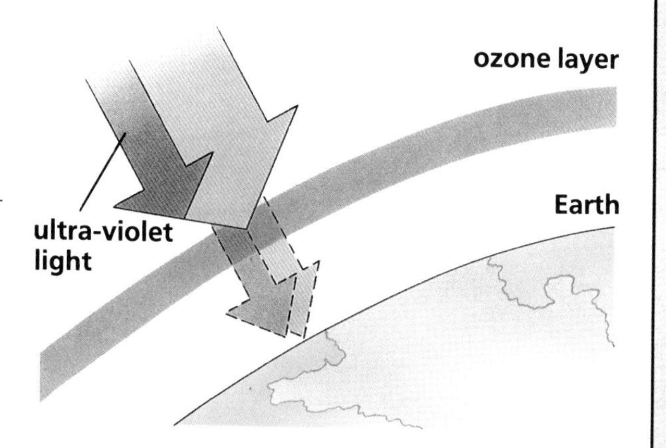

Chemicals called chlorofluorocarbons, CFCs for short, are believed to be mainly responsible for the fall in the amount of ozone. They are used in aerosol sprays, fridges, air-conditioning units and plastic foam cartons. When CFCs are released into the atmosphere, ultra-violet light breaks them down and releases chlorine. This destroys ozone. Look out for ozone-friendly labels.

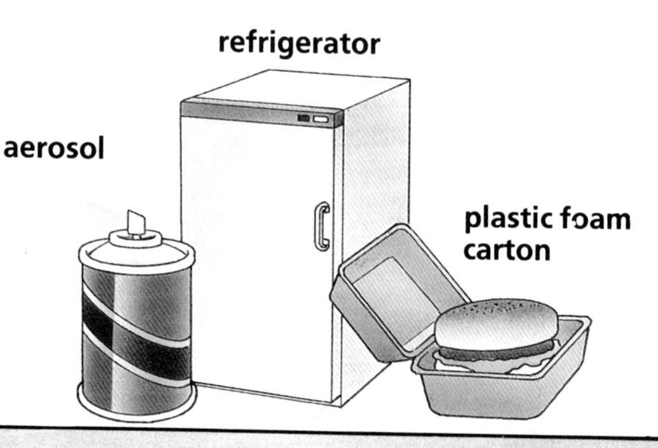

ACID RAIN

Gases like sulphur dioxide and nitrous oxide are released into the air from coal and oil-fired power stations, car exhausts and factories (**1**). These gases rise into the air and are carried by the wind to other areas (**2**). Water vapour in the air combines with the gases to produce acids; they can form acid rain, snow, hail or mist (**3**). Acid rain can destroy plants, turn lakes acid, killing all the fish, and eat into buildings (**4**). Acid rain has spread round the world but Europe and North America are most at risk.

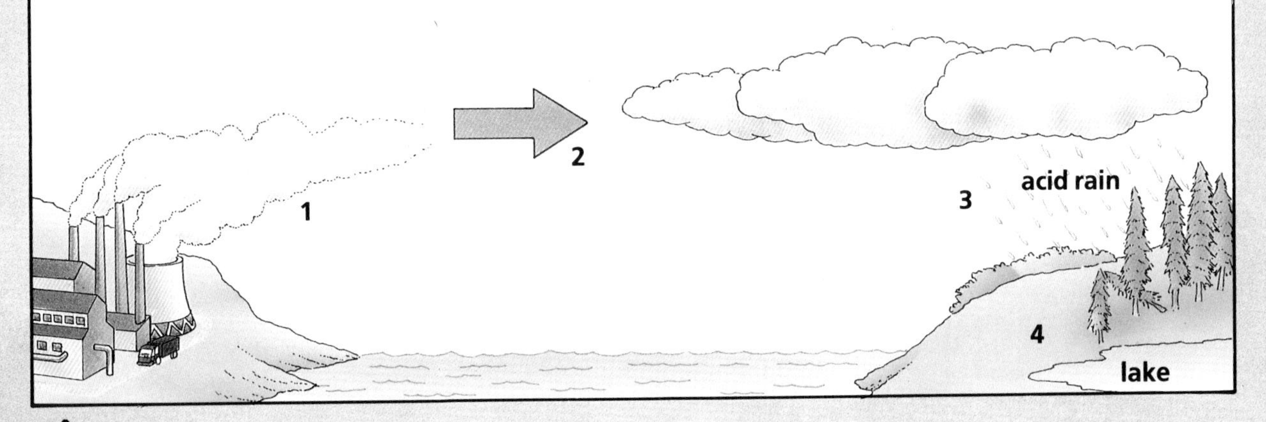

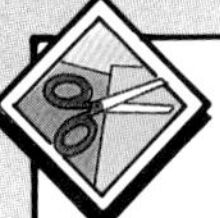

TESTING FOR ACID RAIN

What to do

1 Ask an adult to chop the cabbage into the pan and cover with hot water. Let it stand for 2 or 3 hours. Pour the liquid through a sieve into another container.

2 Dip the blotting paper in the liquid and leave to dry. This is the indicator paper. Dip a piece into acid lemon juice and it should turn red.

3 Dip the indicator paper in some rain water. If it goes red this is acid rain.

USEFUL ADDRESSES AND FURTHER INFORMATION

Friends of the Earth,
26–28 Underwood Street, London N1 7JQ.
Information on changing climates and pollution.

National Centre for Alternative Technology,
Llwyngwern Quarry, Machynlleth, Powys SY20 9AZ.
Information on wind power and solar energy.

Greenpeace,
30–31 Islington Green, London N1 8XE.

Meteorological Office,
London Road, Bracknell, Berks RG12 2SZ.

London Weather Centre,
284/286 High Holborn, London WC1V 7HX.

Buying weather instruments
Many weather instruments can be bought quite cheaply from garden centres, DIY stores and hardware shops. It is well worth looking around to see what is available. You can also buy weather instruments from school suppliers, but they can be expensive. Some have good weather boards for displaying weather readings. Perhaps you could ask your teacher for details. Here are three addresses you could try:

E. J. Arnold, Butterley Street, Leeds LS10 1AX.

Philip Harris Education, Lynn Lane, Shenstone, Lichfield, Staffordshire WS14 OEE.

Geopack, PO Box 23, St. Just, Cornwall TR19 7JS.

WIND CHILL TEMPERATURE CHART

Wind chill is the cooling effect of the wind on the skin. You can calculate this from a chart. Put one finger on the wind speed, say 30 kilometres per hour. Put another finger on the temperature, say −5°C, then move the two fingers along the columns of numbers until they meet. In this example they will meet at −17°C. This means that although the temperature is −5°C it feels like −17°C.

Wind speed (km/h)	Temperature °C							
	0	−5	−10	−15	−20	−25	−30	−35
10	−2	−7	−12	−17	−22	−27	−32	−38
20	−7	−13	−19	−25	−31	−37	−43	−50
30	−11	−17	−24	−31	−37	−44	−50	−57
40	−13	−20	−27	−34	−41	−48	−55	−62
50	−15	−22	−29	−36	−44	−51	−58	−66
60	−16	−23	−31	−38	−45	−53	−60	−68

THE BEAUFORT SCALE

Beaufort number (force)	Wind speed (km/h)	Wind strength	Effect
0	0–1	calm	Smoke rises vertically
1	1–5	light air	Smoke drifts slowly
2	6–11	light breeze	Wind felt on face; leaves rustle
3	12–19	gentle breeze	twigs move continuously; flags unfurl
4	20–29	moderate breeze	Raises dust and loose paper; small branches move
5	30–39	fresh breeze	Small, leafy trees sway; small wavelets on inland water
6	40–50	strong breeze	Large branches sway; telephone wires whistle
7	51–61	near gale	Whole trees sway; difficult to walk against the wind
8	62–74	gale	Twigs break off trees; very difficult to walk; gale warnings
9	75–87	strong gale	Large branches break; chimney pots and roof tiles blow off
10	88–101	storm	Trees uprooted; severe damage to buildings
11	102–117	violent storm	Very widespread damage, usually at sea or in coastal areas
12	over 118	hurricane	Major disaster; countryside devastated

GLOSSARY

Acid raid rain containing dissolved chemicals from power stations etc.

Air pressure the weight of the air pressing down on Earth's surface.

Anemometer instrument used to measure wind speed.

Anticyclone an area of high air pressure. Brings dry weather.

Atmosphere blanket of air round Earth.

Axis an imaginary line drawn through the North and South poles that the Earth spins round once every 24 hours.

Barometer an instrument used to measure atmospheric (air) pressure.

Beaufort scale scale of wind speeds, from 0 (calm) to 12 (hurricane).

Celsius degrees (°C), used to measure temperature.

CFCs (Chlorofluorocarbons) chemicals which are mainly responsible for destroying the ozone layer.

Climate the weather of a particular place, averaged out over a long period.

Cloud a mass of tiny water droplets or ice crystals hanging in the air.

Cold front boundary between two air masses where cold air pushes away warm air. Usually brings rain, followed by colder, brighter weather.

Condensation the process where a gas or vapour cools and becomes a liquid.

Cyclone local name for a hurricane originating in the Indian Ocean.

Depression an area of low air pressure. Cloudy and wet, often with strong winds.

Dew a layer of water which forms on cold surfaces, such as grass, when water vapour in the air condenses.

Equator an imaginary line around the centre of Earth, half way between the North and South poles.

Equinox 21 March or 21 September, when day and night are the same length.

Evaporation the process where a liquid is heated to become a vapour or gas.

Fahrenheit degrees (°F) used to measure temperature.

Fog tiny water droplets hanging in the air near the ground; low cloud.

Frost ice formed when dew freezes.

Greenhouse effect warming of Earth's atmosphere. Caused by a layer of gases which trap the Sun's heat.

Hail pellets of ice in shower clouds.

Humidity amount of water vapour in air.

Hygrometer an instrument used to measure humidity.

Hurricane fierce storm in the Caribbean.

Isobar lines drawn on a weather map, joining points of equal air pressure.

Jet stream a strong wind 5-10 kilometres up in the atmosphere.

Lightning a large discharge of electricity during a thunderstorm, seen as a flash of light.

Meteorology scientific study of the atmosphere and the weather.

Meteorologist scientist who studies the atmosphere and the weather.

Millibar (mb) unit used to measure air pressure.

Orbit the path of Earth round the Sun.

Ozone colourless gas similar to oxygen.

Ozone layer surrounds Earth and reduces the amount of harmful ultra-violet rays from the Sun reaching Earth's surface.

Precipitation water which falls from clouds as rain, snow, hail or sleet.

Rain drops of water falling from clouds, with a diameter greater than 0.5 mm.

Rainbow coloured arc in the sky formed when white light from the Sun is split into its different colours when it passes through raindrops.

Smog fog mixed with polluting chemicals like smoke and exhaust gases.

Snow ice crystals falling from clouds.

Synoptic chart weather map using isobars to show highs, lows and fronts.

Temperature how hot or cold something is.

Thermometer an instrument used to measure temperature.

Thunder the noise caused by the sudden expansion of air heated by a flash of lightning.

Troposphere lowest layer of the atmosphere, directly above the ground, where weather happens.

Ultra-violet radiation invisible radiation from the Sun; this causes suntans, but too much can be harmful.

Warm front boundary between two air masses, where warm air pushes cold air away. Brings rain, followed by warm, cloudy often humid weather.

Water vapour water in the form of a gas in the atmosphere.

Weather the state of the air at a certain time and place. Includes wind, temperature, humidity, precipitation and cloud.

Wind vane an instrument for measuring wind direction.

Wind chill the cooling effect of the wind on the skin.

INDEX